ESSENTIAL SAFETY PROTOCOLS FOR CHEMISTRY LABORATORIES

Authors

Suruchi
Professor, Department of Chemistry
Ramjas College, University of Delhi.

Manish
Dy. General Manager(Chemistry)
Bhartiya Rail Bijlee Company Ltd (NTPC)
Aurangabad, Bihar

Edited by
Yatharth Mehta
Scholar, Department of Chemistry
Ramjas College, University of Delhi

notionpress.com
INDIA • SINGAPORE • MALAYSIA

ISBN
Paperback 979-8-90431-360-9
Hardcase 979-8-90510-923-2

ACKNOWLEDGEMENTS

First and foremost, we thank God for the strength and wisdom he gave us to achieve this goal.

I, Suruchi, would like to thank my Nanaji, who always inspired me to keep moving forward.

I want to thank my husband Manish, who has always supported me immensely and explained many things very clearly while I was writing the book.

A special thanks to my daughter Saanvi, who cooperated with me the most — whenever I thought of giving up, she pushed me forward.

Heartfelt thanks to my students at Ramjas College. Watching them working in the lab sparked the idea to write this book, and our time-to-time discussions turned a messy outline into a real book.

I am grateful to every person who shared their experiences with me regarding lab safety issues.

I express my heartfelt gratitude to my mentor, Prof Rita Kakkar, who inculcated in me the spirit of scientific temperament

I, Manish, would like to express my heartfelt gratitude to my parents for providing me with every opportunity and support to move forward and achieve the best in life.

I sincerely thank NTPC Ltd. for providing me with valuable exposure, facilities, equipment, and an environment that promotes the highest standards of testing and laboratory safety practices.

I would also like to extend my special thanks to my wife, Suruchi, for being my mentor and guide in this first endeavor towards writing a book.

I am deeply grateful to my colleagues and friends at NTPC Ltd. for their constant support, encouragement, and motivation throughout the work on this book.

I would like to thank each and every person who contributed, directly or indirectly, to the writing of this book on laboratory safety.

Finally, we express our sincere gratitude to our families for their patience, understanding, and unwavering support during the long months of writing. We would also like to thank our editor, Mr. Yatharth Mehta, for his valuable editing support and contribution to this work.

Lastly, we express our gratitude to Notion Press for their expertise in publishing the book.

PREFACE

Nobody walks into a laboratory thinking something will go wrong. That, in a way, is part of the problem.

The lab is an exciting place. There is something quietly thrilling about working with substances and instruments that most people will never encounter in their daily lives, about running an experiment that might confirm a hypothesis or, better yet, overturn one. That excitement is worth protecting. And the best way to protect it is to make safety not a burden but a habit, so instinctive that it never gets in the way of the work.

This book grew out of a simple observation: most of the incidents that happen in laboratories are not caused by recklessness. They are caused by gaps. A piece of guidance nobody thought to give. A shortcut that worked a dozen times before the one time it did not. A procedure followed correctly but without any understanding of why, so that when something unexpected happened, nobody knew how to respond. This manual tries to close those gaps, not by burying you in regulations, but by explaining the reasoning behind each practice clearly enough that it sticks.

The guidance here is written for everyone who works in or around a lab setting. If you are a student and this is your first time reading a safety manual, welcome. If you are an experienced researcher who has been doing this for years, there may still be something here worth a second look. Familiarity is comfortable, but it can also be where attention quietly slips. And if you are a safety officer, a supervisor or someone responsible for others, we hope this becomes a resource you can share with confidence.

The topics covered range from the everyday, personal protective equipment, chemical storage, working with equipment, to the less frequent but critical: emergency procedures, spill response, and waste disposal. Each section is written to be clear and direct, easy to find in a hurry, but with enough context to make sense of the bigger picture.

Safety in a laboratory is not just a set of rules posted on a wall. It is an attitude, something that develops over time and spreads through example. We hope this book plays a small part in building that kind of environment wherever you work, one where looking out for yourself and your colleagues is simply part of how things are done.

CONTENTS

CHAPTER
01

SAFETY IN CHEMISTRY LABORATORIES

GENERAL SAFETY AND OPERATIONAL RULES

GENERAL SAFETY

If safety regulations are not strictly followed, chemical laboratories can become some of the most dangerous workplaces. During experiments, students may come into contact with toxic, combustible, or explosive substances that could pose a danger. Mishandling certain chemicals or equipment can cause severe burns, cuts, or other injuries. Ignoring these risks increases the chance of accidents happening [1].

Accidents in chemical labs mainly occur due to carelessness, impatience, unauthorised experiments, or disregard for safety rules. When working or observing in a lab with chemicals, always follow the rules to reduce risks. Avoid working in a lab without understanding its hazards. To lower the chance of accidents, strictly follow these recommendations [2].

Students are expected to exercise extreme caution and make decisions based on the most accurate and up-to-date information available at all times when performing laboratory work. In scientific laboratories, users are routinely exposed to conditions that endanger their lives, such as chemical spills, equipment malfunctions, and exposure to hazardous materials. The vast majority of workplaces have clearly established procedures for regaining control if the situation worsens and for identifying obvious hazards, such as those associated with a typical fire. The purpose of these protocols is to ensure user welfare in the event that the situation deteriorates. However, laboratories pose a wider range of potential dangers, and some of these threats require additional safety measures that are not typically required in other workplaces. For instance, laboratories pose a broader range of potential dangers than other workplaces. Other types of locations typically require fewer safety measures because they are less dangerous. This phenomenon is due to the fact that the majority of the potential hazards identified in laboratories pose a very high risk. The following compendium provides an introduction to secure practices for commonly used laboratory procedures [3].

OPERATIONAL RULES

Stowed items and equipment must not in any way obstruct the fire extinguisher[s], safety equipment, or other emergency response necessities. This comprises all necessary materials. Stairwells, hallways, passageways, corridors, and access points leading to either emergency equipment or exits must be kept dry and free of obstructions at all times. This guideline indicates that no storage, equipment, phone, or other wiring should ever be present in these areas. Paper, wooden containers, and pallets are not permitted to be stowed in the building's hallways or stairwells under any circumstances. This includes the actual structure. It is essential to clear the corridors of all stocks and other materials to prevent obstructions at exits and standard travel routes. This is a prerequisite [4]. Consuming any kind of food or liquid in a laboratory is strictly prohibited. In each and every laboratory, certain office spaces may be assigned for the preparation and consumption of food, and the Safety Committee is contacted regarding this. They are required to maintain a safe distance from all activities taking place in the laboratory at all times. It is not appropriate to bring consumables, chemicals, or instruments from work areas into designated office spaces for the purpose of exchanging these things [5]. It is a severe breach of protocol to keep any food or drink in the laboratory's cold chambers, refrigerators, or freezers at any time. No employee is permitted to perform solitary work in a laboratory or chemical storage facility outside of normal business hours. To request permission to work outside core hours, students and staff must submit a written request to the PI in charge of the project [6]. Animals are not permitted in any portion of the laboratory, with the exception of those being tested and having Animal Experimentation Committee approval to do so. The apparel worn in the laboratory should provide some level of protection against splashes and spills, and be easy to remove in the event of an accident. Non-combustible, flame-resistant, and impermeable aprons offer the highest level of protection at the lowest possible cost. Instead of buttons, laboratory coats and jackets should be equipped with snap closures that allow quick removal if necessary. As part of users' job responsibilities, users must ensure that these jackets are stored appropriately and then remove them before exiting the laboratory. Users are severely discouraged from wearing sandals and other footwear with open toes in the laboratory. Due to the possibility of contamination, this is the case. Laboratory attire must be kept immaculate at all times and replaced whenever it becomes soiled. When working with chemicals or in a facility with a biosafety level of 2, users must always wear lab coats, gloves, closed shoes, and protective eyewear. In addition to that, eyewear or other eye protection must always be worn [7]. At all times, pipetting by mouth is rigorously prohibited.

A. Electrical

The efficient operation of conventional laboratories often calls for the use of a significant amount of available electrical power. As a direct consequence of this, the possibility of electrical problems and hazards has considerably increased. It is absolutely necessary to keep in mind the likelihood that the building catches fire as well as the risk of receiving an electrical shock when having conversations with the people who live there. When developing an efficient electrical safety program for the laboratory, the following components should be given the utmost priority to achieve the desired results: Every single piece of electrical hardware needs to be connected to the ground in some reliable way. It is of the utmost importance to make certain that there is sufficient room for users to move around in the region immediately surrounding the circuit breakers. A label must be affixed to every circuit breaker and fuse, stating whether the device is "on" or "off" and identifying the piece of machinery or section of the room that is receiving power. It is essential that this identifier be incorporated into the code. A prerequisite of the utmost importance is to be in possession of the essential criteria for fuses. To keep each piece of machinery, household appliance, and extension cord [junction box] in pristine shape, they need to be cleaned and dusted on a regular basis. It is against the law to use extension cables instead of permanent wire that has been placed in the correct manner. It is against the law to let wires, cables, or other lines of any type hang freely from the ceiling to the floor in any room or corridor. This includes providing any kind of support for the lines in question. Running wires or cords across metal components such as pipes, ceiling supports, metal shelving, or emergency showers is dangerous and should be avoided. These include, but are not limited to, the things that were described in the clause that came before this one. It is not a good idea to run cables through the floor, wall, or ceiling, or even the frames of the doors or windows. It is not recommended to store this item underneath any other significant objects, including carpets or other heavy goods. It is of the utmost importance to avoid installing cables in locations where they are likely to be frequently mistreated, such as on pavements or in other places, as this might cause the insulation to deteriorate, rendering the cables unusable. Examples of such locations include pavements and other places.

Using multi-outlet receptacles without built-in circuit breakers is unsafe and illegal. This can overload the electrical conductor, causing damage or excess heat. Most portable power strips have outlets that deliver 15 amps each. After making all connections, the staff must ensure the current does not exceed 15 amps. Equipment has manufacturer labels showing its amperage rating. The Electrical Department must handle all facility wiring, splicing, and repairs [8].

The fire extinguisher or extinguishers, safety equipment, and any other supplies required to respond to an emergency must always be easily available, with nothing blocking their path. This is a requirement that must be met at all times. According to the facility's maintenance standards, all stairways, hallways, tunnels, and aisles, in addition to access to any emergency equipment or exits, must be kept dry and clear of obstructions at all times. This applies to both the interior and exterior of the building. This is the case with regard to the inside as well as the exterior of the building. Because of this, storage of any kind, including cables, equipment, phones, and other objects of a similar sort, should never be placed in these areas under any circumstances. It is a violation of the rules to store combustible goods such as paper, wooden crates, or pallets under the stairs or in the hallways. These include, but are not limited to, the things that were discussed earlier in the sentence. It is vital to remove any and all containers and other objects that might be found in the corridors in order to make certain that normal travel routes and exits are not hampered in any manner. It ensures that regular traffic can continue unabated. This is something that can in no way be stopped from happening [9]

When users are meant to be working in the laboratory, it is against the rules of the company to consume any kind of food or drink. This is because the laboratory is a clean environment. This package includes a supply of potable water. It is possible that specific laboratory office spaces will be assigned for the preparation and consumption of food in the event that an agreement is established with the Safety Committee. This would be a positive development. They are obligated to keep a safe distance from any and all activities taking place in the laboratory at all times. Consumables, chemicals, and instruments are not allowed to be moved between the specified work areas and the office spaces under any circumstances. This includes moving the distinct areas to other locations. If a member of the laboratory staff stores food or beverages in the facility's freezers, refrigerators, or cold chambers, they are in breach of a policy that carries severe repercussions [10]. Visits from the general public are welcome at the following times and locations in the offices of NCBS, In Stem, and CCAMP: Our business is open from 8:00 in the morning until 8:00 in the evening, Monday through Saturday. When it comes to the management of the company, these are the times that are of the utmost importance. It is against the law to leave an employee unattended in a laboratory or chemical storage facility outside of normal business hours. This includes shifts that take place overnight as well as those that occur on the weekends. In order to acquire authorization to work outside of regular business hours, members of the institution's staff, as well as students, are required to make a formal request to the PI who is in charge of the project [11].

Any and all animals are strictly prohibited from entering the remaining parts of the building, as per the zero-tolerance policy [12]. The clothing that is worn in the laboratory ought to offer some level of protection against splashes and spills, and it ought to be simple to take off in the event that an accident occurs. Aprons that are non-combustible, flame-resistant, and impermeable offer the best possible level of protection at the lowest cost that is practically possible. Clip closures, as opposed to buttons, should be used on laboratory coats and jackets. These closures make it possible to take off the garments in a brisk and uncomplicated manner. After first securing these garments and taking them off, users are given permission to leave the laboratory. However, the user must first secure them.

Users have to wait until that period of time has passed before they are allowed to leave. The responsibility of completing this assignment comes along with holding this position. On the premises of the laboratory, those working there are not permitted to wear sandals or any other footwear that leaves the toes uncovered. This is done in order to make certain that the working environment is as free from danger as is humanly practicable. Users are required to keep their clothes clean while working in the laboratory and to change them as soon as they begin to show symptoms of wear and tear. When working with chemicals or engaging in other activities that require biosafety level 2, users are required to wear a lab coat, gloves, closed-toe shoes, and protective eyewear at all times. In addition, users must always safeguard their eyes by using glasses or some other type of eye protection. When administering medication to a patient, the tongue should never be used as a conduit for the pipette. The observance of this rule is consistently seen as a matter of the utmost significance [13].

B. Vacuum Operations

A rupture in an evacuated system under vacuum can cause an implosion instead of a detonation because the increased pressure is outside the system. This is because the system is operating under vacuum conditions. This is a consequence of the system-wide decrease in pressure. As a direct result, there is a possibility that the glass may fracture, hazardous chemicals may be released, and a fire may be started. Anyone who uses a Hoover cleaner that has a vacuum should be aware that explosions are always a possibility. Use a vacuum cleaner only on glass items designed for use with it, such as heavy-walled filter flasks, desiccators, and other similar items. Glass that has been fractured, engraved, or otherwise damaged cannot be reused under any circumstances. Before putting anything to use, it should always be inspected for cracks and other flaws. Vacuum glassware must be annealed after being chilled to a temperature equal to or lower than that of liquid nitrogen before use. Due to the fragility of vacuum glassware, this procedure is absolutely essential [14].

All rotating evaporator condensers, receiving containers, and receptacles must be covered with tape or positioned behind safety barriers when operating a vacuum. Either a vacuum pump or a compressor must be used to achieve an adequate level of vacuum for distilling volatile liquids. In both of these circumstances, the use of a chilly trap is essential. This is done to prevent the container from becoming contaminated by escaping solvent vapors. In order for cold traps to be capable of capturing all condensable vapors that may be present in a vacuum system, they must be of sufficient size and temperature. In the absence of such a trap, the exhaust generated by the compression process must be evacuated to the exterior using explosion-proof methods. After a procedure requiring the use of a cold trap has been completed successfully and without incident, it is necessary to purge the system. Any volatile substances that have accumulated in the trap as a result of the evaporation of the coolant may begin to vaporize, leading to an accumulation of pressure that has the potential to cause the device to explode. After the vacuum distillation process has concluded, the vessel in question must be cooled to the same temperature as the ambient air before the venting process can commence. There must always be a barrier or grid-patterned friction film surrounding each vacuum-sealed desiccator. Each of the options presented is acceptable. This allows for the inspection of the contents in the event that the container fractures while preventing the release of broken glass into the surrounding area. Currently, there is an extensive selection of plastic desiccators [such as polycarbonate] available for purchase. Because the use of these desiccators reduces the likelihood of an implosion, it is likely that they are selected [15].

C. Handling Glassware

Handling laboratory glassware requires strict adherence to safety rules to prevent injury from broken glass, chemical spills, or thermal shock. Key rules include inspecting for flaws before use, using appropriate PPE (safety goggles, cut-resistant gloves), handling with two hands to support the weight, and avoiding the use of cracked or damaged items.

Chipped items. Managing glass comes with inherent risks, but adhering to proper safety protocols and handling guidelines can mitigate accidents and prioritize the well-being as follows:

- Transport the glass to the user's side with both hands.
- Avoid transporting glass beneath users' arms or elevated above users' heads.
- Utilize two team members to transport huge sheets of glass.
- Maintain appropriate hand and body positioning to avoid being in the "line of fire" in case of fracture.
- Substitute deteriorated safety equipment prior to doing the task.
- Refrain from standing or walking on shattered glass.
- Do not shatter glass into tiny fragments using the user's foot or hand.
- Avoid grasping shattered glass or attempting to halt its descent; go to a secure area.
- Avoid grasping or seizing glass by the corner of the pane.
- Consistently remain cognizant of users' own and users' colleagues' environments.
- Examine containers and glass before unloading, as this is crucial for mitigating any risks.
- Verify that no team members are present in the vicinity of glass movement.
- Refrain from standing on racks.
- Verify that the glass is secured to the rack prior to transportation.
- Inspect the strapping for deterioration and replace it if required.
- Exercise caution about forklift truck traffic and refrain from positioning oneself in front of any container or glass rack during movement or placement. Confirm the stability of the rack.

- Utilize appropriate safety gear, including hearing protection, gloves, safety footwear, and masks throughout edge deletion and glass-cutting procedures.
- When placing the glass on the floor, position the longer side first, utilizing soft or padded surfaces, such as blankets or other cushioning materials, to ensure the glass's protection.
- It is imperative to have a medical expert or paramedic team readily available when dealing with glass to manage any unexpected emergencies. Furthermore, all users must receive training in providing first aid to their peers.
- Providing thorough training, employing suitable tools and equipment, and following safety protocols and regulations are essential measures to prevent injuries and reduce the risk of glass damage.

Learn and observe the safety and laboratory rules!

When users are working alone in a laboratory, they should NOT engage in any unapproved activities or experiments. Always safeguard users' eyes with the proper equipment when working in the laboratory. In accordance with the law, all Tennesseans are required to adhere to the aforementioned procedure at all times. Prior to use, the user's teacher must approve the safety goggles the user wants to wear to shield the user's eyes from potential chemical leaks. If a user gets chemicals in their eye, go to an eye-wash station and rinse it with water. It is prohibited to have long hair or loose attire when working in a laboratory. Everyone who works in the laboratory is expected to always dress appropriately and adhere to the established dress code. Wearing clothing below the knee must completely conceal users' lower thighs and quadriceps. This applies to users' legs on both sides. At all times, the thighs must be covered with appropriate apparel. Users have the option of changing into a lab coat or an apron that covers users' legs from the thighs to the knees if the clothing users are presently wearing does not meet the requirements. Users must always wear stockings and shoes, and the soles of their footwear must be closed. This establishment prohibits sandals, open-toed shoes, and shoes without backs, sling backs, loafers, and open-toed sling backs. It is essential to be familiar with the locations of fire extinguishers, fire blankets, safety showers, eye cleansing devices, and first aid supplies, as well as their proper usage. Before making any chemical purchases, the user must thoroughly examine the labels on the containers containing the reagents. In a chemistry laboratory, it is against the regulations to consume any type of food or drink, as well as to smoke cigarettes or consume any liquid. Verify that users' hands are thoroughly clean before leaving the laboratory, and only then should the user proceed to depart. When working with potentially hazardous or irritating vapors, the user must use the pollution shelters provided. The use of a buckle vacuum to fill pipettes is prohibited under all circumstances. Never force glass tubing through latex or cork stoppers without first lubricating the transition points. This action can directly cause the glass tube to shatter into fragments. Under no circumstances should the exposed end of the testing apparatus be directed at the user or another individual. This arrangement places the user in an extremely vulnerable position [19].

Under no circumstances should excessively concentrated acids be allowed to come into contact with water. After using the boiler, ensure that the flame has been extinguished and that users are familiar with the proper procedures for starting and operating a laboratory burner. Before using the stove, the user must remove all items that could catch fire from the kitchen. Regardless of whether they are holding liquid or solid refuse, debris containers must always be utilized appropriately. This is the case regardless of whether or not they contain detritus. In no manner whatsoever is it permissible to place chemicals on the pan of the scale. Such behavior is a flagrant disregard for norms. When weighing a chemical with a balance, it is essential to always use the correct measuring container. Rapidly dumping chemicals onto a scale is a practice that is unacceptable under all circumstances. Unless instructed otherwise by the user's instructor, the user should never return unused chemicals to their original storage containers. After removing the reagents from the containers, the user must replace the containers' closures, coverings, and stoppers in a secure manner. Make it a habit to thoroughly cleanse the spatulas before and after using them in containers containing toxic substances. Regardless of the severity, any accident or injury should be reported to the instructor as soon as possible. If additional items were set up on the laboratory workstations, they could potentially disrupt the procedure, leading to confusion and increasing the risk of accidents during experiments. Before entering the laboratory, users must store all of their personal belongings, including books, in one of the bookcases strategically placed throughout the facility [20].

The accumulation of trash and its eventual disposal require immediate attention and care. Ensure that users have thoroughly cleaned and dried their work area before leaving the laboratory. Verify that the gas, water, vacuum cleaner, and air valves are all completely closed and in the correct position. Do not hesitate to approach the user's instructor at any time with any inquiries or concerns the user may have; they are always available to help. Never hesitate to ask questions, especially if users are unsure how to do something. Ask questions without ever displaying reluctance. Before proceeding, the user must first ensure that they have a comprehensive understanding of the instructions [21].

In order to meet these requirements, it was necessary to conduct an experiment with diluted sulfuric acid. The lecturer instructed the students that acid must always be diluted with water and that the procedure should never be performed in reverse. In addition, the teacher instructed the class to dilute the sulfuric acid concentration with water in a ratio of one part acid to four parts water. Furthermore, the instructor emphasized that when combining substances, acid should always be added to water, and water should never be added to acid. It is essential to keep this in mind when combining substances. Because acid neutralizes water and water neutralizes acid, it is true that this rule is the case. John failed to follow the instructions, so he proceeded to combine the acid and water. As a result of the explosion, the beaker reached an abnormally high temperature, a mist formed over the solution, and some of the solution overflowed onto his and his partner's skin. The detonation produced an almost inaudible noise [22].

Hazards and Risks

With the aid of this manual, users will learn how to safely carry out their responsibilities in the laboratory, not only for themselves but also for others. To ensure users' safety, regardless of where users travel, users must bear in mind the following four factors:

- Be mindful of prospective risks.
- Conduct a risk assessment so that the user can identify the potential dangers involved.
- Ensure that the necessary precautions are taken to reduce the likelihood that potential hazards cause damage.
- Prepare for emergencies. [87]

RAMP is a mnemonic device that can be used to help remember the actions that must be taken: they are identified, evaluated, minimized, and planned for. This topic is discussed in a variety of contexts throughout the entire book. Let's examine the reasoning that led to the creation of these four stages, bearing in mind the multitude of safety considerations that were taken into account.

The essence of the concept of security is the absence of any prospect of experiencing physical harm or property loss. If the user wants to keep everyone secure and ensure no one is injured, all parties may have to contribute. If users make safety a priority in the college lab, they avoid situations that could harm their health. Even though conducting chemical experiments in a laboratory may appear to be a simple task, it is frequently easier to become "caught up" in the procedures and neglect to conduct the experiment in a safe manner. This is especially true if the task is being performed by an inexperienced individual. This is because it is occasionally easier to become "caught up" in routines. The first portion of each chapter in this book is devoted to a discussion of a laboratory incident that could have had potentially hazardous consequences. Everyone who has experienced these circumstances is "just like a user." They had received training in safety precautions, but at the time of the accident, they gave no consideration to safety. Additionally, users can eventually find themselves in a similar circumstance. Let's review some of the fundamental concepts associated with hazards and risks, as well as some strategies that can be used to prevent accidents at research institutions. A hazard is any event that increases the likelihood that someone is injured or loses something. When we say that something has the potential to be detrimental or destructive, we mean that it is capable of acting in such a manner. There is a possibility that a very large number of distinct compounds, each of which possesses its own unique intrinsic dangers, have not changed over time. In actuality, the objective of the safety procedure is to lessen, manage, or gain control over these risks in some other manner, and there are a variety of methods for achieving this objective. If the user reads this book and pays close attention in the user's chemistry classes, the user learns a great deal about the myriad forms that chemicals can take and the dangers presented by the compounds the user studies [23].

The term "risk" indicates the possibility of getting harmed as a direct result of being exposed to a hazardous or risky circumstance. In addition to the inherent risk presented by a chemical, a number of other factors influence the degree of danger it poses. For instance, the quantity of a chemical, its physical state [whether it is a gas, liquid, or solid], and the manner in which it is handled can all influence the level of danger it poses [24].

When we talk about being "exposed" to a hazard or chemical, we mean to come into contact with them in a manner that can cause harm or injury. One dose consists of a single application of a drug to the epidermis, one single inhalation of the substance, or one single absorption of the chemical via food. There is the prospect of being exposed to the elements in the event of a fire or explosion. Dose, duration of exposure, and mode of exposure are the three most crucial characteristics to consider. Each of these variables has the potential to significantly impact the severity of the injury. Keep in mind that users are being led into potentially hazardous situations on purpose to minimize or eliminate the dangers users face. This is the most significant conclusion that can be drawn from all of this information [25].

Dangers are a Necessary Part of Our World!

Although a variety of items pose a variety of potential risks or hazards, we have acquired the knowledge and skills necessary to utilize them safely and effectively. In many instances, the beneficial properties of compounds also contribute to their toxicity [please see Figure 1 of the attached document for an illustration of this concept]. Even if something poses a threat to our health, that does not necessarily mean we should avoid using it. In fact, our very existence, as well as our ability to experience a certain level of comfort, is frequently dependent on the use of hazardous chemicals and machines. This is true with regard to both our ability to survive and our capacity to do so in comfort. Let's examine benzene as an example of one of the most pervasive chemical hazards that the majority of us face every day: benzene exposure. Depending on the circumstances, gasoline is highly combustible and has the potential to ignite or even erupt when ignited. This is only possible because we are aware of petrol's hazardous properties, have assessed how we may be exposed to them, have developed techniques to effectively minimize or manage them, and have learned how to both plan for and respond to situations involving petrol. We frequently visit petrol stations with subterranean storage tanks holding thousands of gallons of fuel; we pump gallons of fuel into our own vehicles; and we frequently travel with gallons of fuel stored in the petrol tank of a nearby parked vehicle. Because we are now aware of how to safely handle this potentially dangerous substance, we are no longer at risk. Do substances or other things exist in our environment that, despite their many useful applications, possess potentially hazardous properties and could place us in danger if we treat them irresponsibly, fail to effectively manage them, or make improper use of them? If so, what are their characteristics? Consequently, what are these distinguishing characteristics, and how can we identify them? Consequently, what exactly are these characteristics? Do users believe that the majority of 21st-century residents would embrace the introduction of a high-pressure, explosive chemical that, if ingested in sufficient quantities, could prevent a person from breathing oxygen? The feedback we've received has been almost exclusively negative! However, a significant number of us are responsible for this because we use natural gas. Which elements comprise the characteristics of the electric current? Which perils should we be most concerned about? How does the risk become evident when viewed on a global scale? All members of the security community have acknowledged that ubiquitous threats pose less of an overall risk than less pervasive threats. Even if the preponderance of laboratory threats initially appears to be "uncommon," it should be noted that they are not significantly more dangerous than the everyday threats we face [26].

At its most fundamental level, the concept of security refers to a situation in which there is absolutely no danger of suffering any type of physical damage or material loss. If they want to ensure that no one is injured and that everyone remains secure, both parties need to make an effort and work together. If users determine that safety is their top priority in the college laboratory, they avoid situations that could place their physical health or other aspects of their health at risk. For instance, a user may determine that avoiding situations in which they could be exposed to hazardous chemicals is the most important thing. This includes avoiding situations where the user may be exposed to potentially harmful substances. Even though conducting chemical experiments in a laboratory may appear to be a straightforward process, it is frequently easier to become "caught up" in the procedures and forget to conduct the experiment in a safe manner. This could have extremely negative consequences, especially if the individual performing the task has little to no prior experience working in the relevant industry. This is due to the fact that it is sometimes easy to become "caught up" in routines, which is why this is the case. This is owing to the ease with which routines can be "caught up." The first few paragraphs of each chapter in this book are devoted to a description of a laboratory event that may or may not have had potentially detrimental consequences. Everyone else who has experienced something analogous to what users are currently experiencing has been described as "just like the user." Despite having received training in the proper safety procedures, they did not consider safety at any stage during the accident. At some point in the not-too-distant future, users encounter a situation strikingly identical to the one users are currently experiencing.

Let's examine some of the fundamental concepts associated with risks and hazards, as well as some of the solutions that can be implemented to prevent unintended outcomes at research facilities. A risk factor is any condition or occurrence that increases the likelihood that a person can sustain an injury or lose property. When we say that something is capable of being harmful or destructive, we mean that, if given the opportunity, it could behave in a way that is harmful or destructive. This is what we mean when we say that something has the potential to be detrimental or negative. It is possible that there are multiple compounds, each with its own inherent dangers, and that these dangers have not changed over time. This is a possible outcome. This is one of several possible outcomes. In actuality, the purpose of the safety procedure is to reduce, manage, or gain control of these risks in some other manner, and there are numerous ways to accomplish this objective. In addition, the objective of the safety approach is to reduce the catastrophe risk to an acceptable level. If the user reads this book and pays close attention in the user's chemistry classes, the user will learn a great deal about the various forms compounds can take and the dangers associated with the substances the user studies. In addition, users learn a great deal about the various interactions between compounds. If the user peruses this book, the user will gain an understanding of the various forms that chemicals can take.

The noun "risk" should be used when referring to the possibility of sustaining an injury as a direct result of being subjected to a hazardous or risky state. In addition to a chemical's inherent peril, a number of other factors are taken into account when determining its potential impact on human health and safety. Despite the chemical's inherent danger, this situation has arisen. For instance, the quantity of the chemical, its physical state [i.e., whether it is a gas, liquid, or solid], and the manner in which it is handled can all affect the level of danger it poses.

When we say that a person or object has been "exposed" to a risk or chemical, we mean that they have come into contact with it in a manner that could put them at risk. As a method of administration, one application to the epidermis, one inhalation of the drug, or absorption of the chemical through nutrition could all be considered one dose. Considered interchangeable with one another, these three processes are listed here. There is a good possibility that users are exposed to the elements in the event of a fire or an explosion. Considered to be of the utmost importance are the doses, the duration of exposure, and the route of administration. Each of these factors has the potential to have a sizable effect on the total quantity of damage. Keep in mind that users are being intentionally led into potentially hazardous situations so that the risks users face can be minimized or even eliminated. This is done so that the risks the user confronts can be minimized or even eliminated. This is done to provide users with complete protection against the hazards to which they are susceptible. This is the one piece of information that should always be at the forefront of users' minds, as it is the single most important takeaway from all of this [27].

FIGURE 1: A quantity of petrol measuring 1.5 liters. Numerous potentially dangerous substances exist in and around our residences, but because we are so accustomed to them, we may perceive them as more hazardous in a laboratory.

Safety from the Experts

The surrounding environment comprises the facilities, location, equipment, procedures, and standards. The individual, which comprises their attitude, beliefs, personality, knowledge, skills, and abilities, and the behavior, which comprises both secure and dangerous practices. Facilities, location, equipment, procedures, and standards constitute the encompassing environment. The individual possesses all of these characteristics. The external environment, the individual, and the behavior all contribute to the overall safety of a setting, according to Geller's research. Due to their interdependence, each of these elements can exert influence over the others. Consider a time when a user witnessed someone lose their equilibrium due to something in their environment, such as a misplaced object, that caused them to lose their balance. When a user tells themselves, "I need to be more cautious around objects on the floor that I could trip over," the user must consider a personal aspect of the situation. If users are walking down an aisle and observe an incorrectly placed wastepaper receptacle, users should move it to the correct location so that it is out of the way and not visible to others. If one wishes to maintain their current level of safety, it is crucial that careful consideration be given to each of these three factors. This is due to the fact that each of these characteristics has a substantial impact on the risk to which an individual is exposed. This is the case because if the laboratory where users do their college work is safe, and if users observe other students and professors working safely by adhering to safety regulations and safe practices, users start to believe that "laboratory safety is important, and I want to be safe as well." If the laboratory where users do their college work is safe, and if users observe other students and professors working safely by adhering to safety rules and safe practices, then users are eligible for this scholarship. User receives a passing grade if the college laboratory in which the user completes the user's assignments is safe and if the user observes other students and faculty members adhering to safety standards and safe practices [28].

How Do We Learn Safety?

The collection of empirical evidence is a crucial aspect of the safety research process. This study demonstrates that we routinely improve our safety by learning from our past mistakes and incidents. The past events can provide valuable lessons; however, if users choose to learn from the same mistakes we made, their survival may be at risk! Provided the opportunity, the overwhelming majority of us would avoid potentially fatal situations such as fires, explosions, chemical exposures, and other events that pose a threat to our health or even our lives. Instead, we must educate ourselves on the safety rules that have been established because of the negative experiences of others in the past.

If we are more cautious and persistent in our efforts to incorporate health and safety concerns into our daily decisions, we are able to reduce the likelihood of negative outcomes. This can be achieved by emphasizing safety and taking all necessary precautions to eliminate the risk of accidents, particularly in the laboratory, such as conducting regular safety drills and ensuring proper training for all personnel. Unfortunately, security rarely receives positive feedback, while safety issues often receive negative feedback. This is a substantial distinction. There is something about this situation that must be altered. If users place themselves or others in unnecessary danger by taking unnecessary risks, there is a high chance that someone will be injured as a result of users' actions. This could be a user or another individual [29].

It Is Crucial to Take Every Available Precaution to Minimize the Risk of Something Going Wrong!

To ensure that working conditions in the laboratory are as risk-free as possible for all users, exposure risks must be minimized or eliminated to the utmost extent possible. Every day, we make it a top priority to safeguard ourselves by strictly adhering to the safety protocols that we have been instructed to follow in order to significantly reduce the likelihood of being involved in an accident. This approach significantly lowers the chances of sustaining any injuries. Table 1 outlines the most significant risk factors associated with these risks, as well as the tested preventative measures that can be taken to mitigate their impact. If we were in a car accident and donned seat belts, for instance, the likelihood that we would sustain life-threatening injuries would be significantly reduced. This reduction is due to the fact that seat belts were designed to prevent specific categories of injuries; this is the case. It is important to emphasize the word "less likely" in this context because wearing a seat belt does not guarantee that the user will not be injured in the event of an accident; however, it does increase the user's chances of surviving the collision.ote that donning a seat belt does not guarantee that the user will not sustain injuries in the event of an accident. It is essential to emphasize the phrase "less likely" in this context, as wearing a seat belt reduces the likelihood of injury in an accident but does not guarantee protection. However, donning a seat belt reduces the likelihood of sustaining injuries. And if users also adhere to other public safety standards, such as the speed limit and defensive driving techniques, they are much more beneficial to users in reducing the risk of accidents and injuries on the road [30]. For example, users are required to wear splash goggles while working in the laboratory to reduce the risk of chemical splashes causing eye injuries. The goggles are not a guarantee that the user cannot sustain an eye injury; however, prior research has shown that donning splash goggles significantly reduces the likelihood that users' eyes may sustain an injury. Taking additional precautions, such as covering one's head with a hood or face shield and wearing protective gloves, reduces the risk of injury even further.

Common Risk	Risk Factors	Risk Reduction Action
Cardiovascular disease, death from- heart disease and stroke, 1st and 3rd leading causes of death	High blood pressure, high blood cholesterol	A 12-13 point reduction in blood pressure decrease cardiovascular disease by 25%; a 10% decrease in total blood cholesterol decreases coronary disease up to 30%
Lung cancer, death from	Cigarette smoking-early death; 22 times higher for male smoker; 12 times higher for female smokers	Stop cigarette smoking
Vehicle accidents, death from	No seat belt use; alcohol-impaired driving	Seat belt use laws; seat enforcement laws; sobriety checkpoints; reducing blood alcohol concentration to 0.08%; minimum legal drinking age laws

Table 1: Common Risks, Risk Factors, and Risk Reduction Actions

The vast majority of incidents are caused by reckless behavior, specifically the taking of unnecessary risks.

A significant number of preventable fatalities and injuries are attributable to the practice of engaging in dangerous behavior, which includes taking excessive risks. These catastrophes could have been avoided. Infractions of safety principles, safety standards, or security procedures constitute one of the seven categories of risk that are entirely unnecessary risks. If the user can identify a method to avoid taking unnecessary risks, there is a favorable chance that the user will be able to prevent or at least reduce the likelihood of something negative occurring.

A person may intentionally disobey the law, act in a manner that is unknowingly influenced by their past experiences, or be oblivious to the potential risks. These are all examples of deliberate disobedience. These three options are all valid considerations and have a high likelihood of occurring. A person is considered to be taking unnecessary risks when engaging in activities such as speeding, not wearing a seatbelt, smoking cigarettes, overeating, not exercising, using illegal substances, and avoiding safety regulations at work, such as choosing not to wear required protective equipment. Other hazardous behaviors include driving without a seatbelt, insufficient physical activity, and illegal substance use. Inappropriate laboratory conduct includes not wearing safety glasses, consuming food or alcohol in the lab, donning inappropriate attire, and taking excessive risks with hazards. Inappropriate conduct also includes not donning a face mask or respirator. In addition, the laboratory is a potentially hazardous environment. Inappropriate behavior includes not only inappropriate actions but also inappropriate apparel and unwarranted risk-taking with potential dangers.

Potentially hazardous situations are not always simple to avoid. Despite the fact that our human nature prompts us to act in ways that are convenient, comfortable, or expedient, there are numerous instances in which it is necessary to act contrary to these inclinations. Some individuals have realized through personal experience that they can save both time and resources by taking detours and that, in many cases, they can do so without negative consequences. This realization has occurred as a result of the understanding that detours can save both time and resources. People are able to get away with violating safety standards, and as a result, they begin to believe that it is acceptable to engage in hazardous or harmful behavior because they can. They do this because they can get away with it. A pattern of destructive behavior is developing rapidly as a result. Due to the fact that people are becoming less vigilant and violating an increasing number of rules and regulations, the likelihood of them being involved in an incident is directly rising as a result of this trend [31].

CHAPTER

02

SAFETY RESPONSIBILITY IN CHEMISTRY LABORATORY

Introduction

To develop a culture of safety and security, one must be unwaveringly committed to upholding high standards on all fronts. One can only expect to achieve their goal in this manner. From the highest levels of institutional leadership, this commitment must extend all the way down to the most fundamental aspects of laboratory operations. Until then, however, it is not possible to construct a culture of safety and security. We provide an overview of a strategy for implementing safety and security measures in smaller chemical research facilities in the following chapter.

After installing a management system for safety and security, the operational aspects of the laboratory become more functional, leading to improved compliance with safety regulations and a reduction in accidents. This technology can also aid in predicting and preventing events that have the potential to cause damage to people and the environment, such as illnesses and injuries. A natural catastrophe is an example of such an occurrence. The scope of the organization, the activities in which it engages, and the risks and conditions unique to its business operations all have an influence on the manner in which the various components of this framework are implemented [32], which can include strategies for mitigating the impact of natural catastrophes and enhancing safety measures for both people and the responsibility for environment

Laboratory Safety and Security

When it comes to establishing and enforcing safe and secure procedures, various individuals within an organization have different tasks and responsibilities. The best method for individuals of all ranks to demonstrate their commitment is by setting a good example for others to follow, such as adhering to safety protocols and actively participating in training sessions [33].

FIGURE 2

Leaders

The ultimate responsibility for ensuring the safety and security of a particular location rests with the institution's administrator and the heads of the numerous departments that comprise it. In the not-too-distant future, it is conceivable that employers will be required by law to ensure that their users are able to carry out their duties in a safe environment. The extent to which executives are able to persuade users to participate in a safety and security program may also influence the ultimate outcome, as higher participation can lead to a more effective implementation of safety measures and a greater overall sense of security among staff. Even if a project is well-designed, if upper management does not make it a priority, it is neglected by the staff, which can lead to increased safety risks and a failure to meet legal requirements for a safe working environment.

Chemical Safety and Security Officers [CSSOs]

All organizations strongly recommend employing a chemical safety and security officer, also known as a CSSO. This group promotes a coordinated effort to manage all aspects of safety and provides guidance to users at all levels. The CSSO is the organization in command of this endeavor. The CSSO should have the knowledge, responsibility, and authority to design and implement a competent safety and security management system. There is a possibility that the function is held concurrently by multiple individuals, who may then divide the responsibilities according to the situation's needs [34], which could lead to challenges in coordination and accountability if not managed effectively.

RESPONSIBILITIES OF THE CSSO

Adhering to laboratory chemical policies and ensuring compliance with applicable regulations as required. Developing and implementing an integrated safety and security program covering the entire life cycle of all laboratory chemicals; providing assistance in the areas of purchasing, storing, using, and disposing of waste at the laboratory level; operating a waste management program for the offsite disposal of waste, if necessary; the program includes the receipt of waste, transportation, and final disposal of the material through the services of commercial vendors.

- Registering orders for laboratory chemicals.
- Receiving chemical deliveries and maintaining an accurate inventory.
- Conducting audits and inspections to guarantee compliance.
- At least once per year, audit the inventory records and cabinet security.

In the event of disobedience, the authorization to use laboratory substances is revoked. Keeping exhaustive records of standard operating procedures [SOPs] for a program in an easily accessible format so that they can be retrieved, disseminated, and inspected as follows:

- Managing chemical-related incidents and conducting investigations [such as spillage, lost chemicals, injuries, etc.
- Providing managers, supervisors, and users with training on how to construct appropriate SOPs and adhere to the safety program [35].

Organization for Environmental, Health, and Safety

In addition to other experts in the fields of chemical safety, engineering, occupational medicine, fire safety, toxicology, and possibly even other fields, some of the larger educational facilities also have an environmental health and safety office with one or more CSSOs on staff. An office of this kind would contribute to the formulation of policies and the promotion of laboratory safety standards. Issues pertaining to hazardous waste, reviews of accidents, inspections and audits, compliance monitoring, training, recordkeeping, and emergency response are frequently included in its scope of responsibility [36].

Personnel in Charge of the Laboratory, Including Supervisors and Teachers

The CSSO and the laboratory manager are directly responsible for the safety and security management program. Laboratory instructors bear direct responsibility for the actions of students in a laboratory setting. The responsibility of encouraging a culture of safety and educating students and other workers on the skills necessary to handle chemicals safely falls on the shoulders of the instructors [37].

RESPONSIBILITIES OF THE LABORATORY MANAGER OR SUPERVISOR

1. Ensure that laboratory personnel receive training on general chemical safety and security measures.
2. Ensure that laboratory users are aware of how to interact with chemicals safely. Offer chemical- and procedure-specific training, including SOP [standard operating procedure] creation and review instruction. Providing laboratory personnel with the necessary engineering controls and personal protective equipment [PPE] to ensure safe chemical interaction in the workplace.

3. Ensure that the laboratory has adequate safety precautions in place for the chemical substances.
4. Establish appropriate protection and safety standards, including consideration of health and safety in performance evaluations.
5. Evaluate and grant ultimate approval for laboratory chemical-related activities. [38]

Workers and Students

A user, along with other laboratory personnel, performs the actual task. However, they depend on the guidance and instruction of their supervisors and instructors. They are responsible for ensuring the proper and secure handling of the substances they employ. Anyone operating in a laboratory, whether a student or an employee, is responsible for adhering to all safety and security procedures to protect themselves and others [39].

RESPONSIBILITIES OF STUDENTS AND WORKERS

1. Participating in instruction regarding the laboratory's safety.
2. Reviewing written instructions and then adhering to written procedures.
3. Prior to beginning work with a new chemical or technique, ensure that users have a thorough comprehension of all potential dangers and safety and protection procedures. SOPs may be reviewed in addition to being developed and approved.
4. If users are uncertain about the hazards, they should consult with the laboratory director or the CSSO.
5. When necessary, implement engineering controls and employ personal protective equipment.
6. Notifying laboratory management immediately of any accidents, security concerns, or potentially hazardous chemical exposures.
7. Creating comprehensive documentation of the required procedures when working with extremely hazardous chemicals or apparatus. Adapting procedures to changing circumstances [40]

Ten Steps to Create an Effective Laboratory Chemical Safety and Security Management System

For a safety and security management system to be effective, the commitment of institutional leaders is one of its most essential components. Leadership is accountable for initiating the development of a strategy and delegating its execution to others.

Establish a committee for institutional safety and security oversight and appoint a chief security and safety officer.

The institution's most senior executive is responsible for establishing a committee that oversees chemical safety and security. The committee must consist of representatives from all affected groups and organizational levels. The senior leaders should be the direct point of contact for the committee, and they should provide the essential financial and administrative support.

In addition, the organization should designate at least one CSSO to oversee the safety and security management program. It is essential to identify the responsibilities and accountability of the CSSO and then communicate this information to the CSSO, as well as to the institution's leadership, laboratory managers, staff, and students. [For additional information on the CSSO's responsibilities, please refer to Section 2.2.2.] In addition to the necessary authority, a CSSO must have sufficient time and resources to fulfill his or her responsibilities. The CSSO should have unfettered access to senior officials, who are ultimately accountable to the public. If the Chief Security and Safety Officer [CSSO] does not have direct access to senior-level authorities, the institution must provide senior leadership with alternative reporting mechanisms, such as regular safety audits and incident reports, to ensure that safety concerns are addressed promptly and effectively. Create a policy for chemical safety and protection.

Those in control of an organization should formulate an official policy to specify and record a chemical safety and security management system. A documented policy statement demonstrates the institution's support and establishes community expectations. In the policy, the following objectives should be stated: Prevent or mitigate human and economic losses caused by accidents, adverse occupational exposures, and environmental events; Integrate safety and security considerations into all phases of operations; achieve and maintain compliance with laws and regulations; and continue to enhance efficacy.

The policy statement should be communicated to the institution's personnel, posted in a conspicuous location, and reviewed and updated as necessary.

It is essential to establish administrative controls and procedures for performance evaluation.

Administrative controls specify the responsibilities of personnel, as well as the rules and procedures that pertain specifically to safety and security. The administrative controls in place must also include mechanisms for managing and responding to change, such as the introduction of new processes, technologies, legal requirements, users, and organizations.

These documents should also define the responsibilities of laboratory students, users, managers, institutional leaders, contractors, emergency service providers, and visitors. The CSSO should devise general safety rules, procedures for laboratory housekeeping, manuals for the use of materials and equipment, and other documents to convey expectations to all laboratory personnel.

Evaluation of the safety and security of laboratory procedures should be an integral element of daily operations. For example, at the beginning of each meeting of a department or group, have a "safety moment" in which users discuss a daily activity, the safety or security issues that the activity entails, and what can be done to prevent potential incidents from occurring.

Determine which situations are especially perilous, such as handling hazardous materials or operating complex machinery, and take steps to mitigate them.

Managers, principal investigators, lead researchers, team leaders, and administrators must all actively manage the safety and security of their laboratories. Conduct a preliminary status review to evaluate the scope, adequacy, and application of safety procedures. Utilize the status review as the foundation for the development of a safety and security program and as a tool for determining improvement priorities. Conduct a risk assessment to determine whether or not the current control mechanisms are adequate, to establish a hierarchy of requirements, and to include corrective actions based on their significance and the available resources. [41]

Responsibility For Accident Prevention

Accident prevention is a shared responsibility that requires the active participation of all personnel in the laboratory. Accident prevention is everyone's responsibility, but particularly that of the individual in charge of performing the laboratory procedure. Accidents are frequently the result of three factors: irresponsibility, a lack of common sense, and carelessness.

Failing to follow the instructions and committing errors. Users run the risk of falling prey to their errors. User risk falling victim to the mistakes made by another student. Be grateful if a classmate tells the user that the user is doing the task wrong and that they may have just saved the user's life. On the other hand, if a user observes someone else committing an error, the user should correct them. The users' instructor is also responsible for sustaining a risk-free environment in the laboratory; please inform them of any potentially hazardous behavior. [42]2]

Participate actively in the disaster prevention activities by engaging in them, such as attending safety drills and reporting unsafe conditions to the users' instructor. Observe the following standards as a general rule:

Pay careful attention to all safety instructions. Never engage in practical jokes or horseplay in a laboratory that handles chemicals. Familiarize users with the location of safety equipment and amenities, including exits, safety showers, and eyewash facilities. Additionally, users should discover how to use them properly.

Before undertaking any type of laboratory work, familiarize oneself with the hazards posed by the chemical in question. Familiarize users with the device's and individuals' potential hazards and which actions to perform and which to avoid. Observe all necessary safety precautions. [43]

PERSONAL PROTECTION

Eye Protection

Everyone in the laboratory, including visitors, is compelled to wear goggles designed to protect against chemical splashes at all times, even when not performing a chemical operation. Normal prescription eyeglasses may exceed the United States Food and Drug Administration [FDA] standards for impact resistance, but they do not provide sufficient eye protection for laboratory use. Never entrust users' safety to this type of eyewear when working in the laboratory. The environmental conditions determine the type of eye protection that is necessary. Any environment where an accidental chemical discharge is plausible does not allow contact lenses to provide adequate protection. Protecting one's eyes involves donning safety goggles similar to those shown in the illustration. The user must always wear safety eyewear in the laboratory, regardless of whether or not the user wears contact lenses. Use the proper laboratory bench shields when working at lower pressures or when there is the possibility of dynamic reactions, implosions, or excessive spillage. Protect users' eyes with safety lenses and users' neck and ears with a face shield that is sufficiently large and long. [44]

Clothing

The apparel worn in the laboratory should provide protection from splashes and spills, be simple to remove in the event of an accident, and be reasonably practical. If the user prefers to wear a lab coat or jacket instead of an apron, the coat or jacket should have clasp closures rather than buttons for easy removal.

When working in the laboratory, users should wear shoes with leather or a polymeric leather substitute on the upper. Do not stroll or carry bags while users are there. It is inappropriate to wear shoes with high heels, exposed toes, uppers made of cloth or woven leather stripes, or any other material typically associated with women. Shorts, cut-offs, and miniskirts expose users' skin unnecessarily to potential corrosives but are otherwise innocuous. Reduce the length and looseness of users' garments. In the laboratory, users should not wear jewelry, including rings, bracelets, and wristwatches. Chemical vapors and leaks pose a threat and can destroy jewelry.

The chemical seepage that can occur between the epidermis and jewelry can expose the skin to corrosive substances. When users wear jewelry, they are more likely to come into contact with electrical sources, which can result in a potentially lethal electrical discharge. In addition, jewelry can become entangled in machinery, which can result in accidents, such as severe injuries or even fatalities, particularly in industrial settings where heavy equipment is used. [45]

Gloves

Gloves is a necessary component of adequate personal protection. Users' instructors may require users to utilize them when appropriate, especially during activities that involve exposure to hazardous materials or when maintaining hygiene standards is crucial. In addition to latex, neoprene, and butyl rubber gloves, a wide variety of other gloves are available for purchase. There are various variations of mittens, each with a unique gauntlet length. Some gauntlets encompass the entire arm, while others cover only the forearm, and others end at the wrist. People who have a latex allergy should avoid donning latex gloves. Do not rely on clothing and leather gloves to protect the user from hazardous substances. These materials may provide some protection against hot and chilly objects. The history of leather gloves makes them more susceptible to contamination than linen gloves, as leather can absorb and retain hazardous substances more easily than the non-porous fabric of linen gloves. linen gloves.

Make sure users are wearing mittens correctly. Before putting on users' mittens, users should always examine them to ensure that there are no cracks or holes. Remove users' gloves prior to departing the work area and handling items such as telephones, door handles, writing instruments, laboratory notebooks, and textbooks. This prevents the user from accidentally distributing the chemical. [46]

CHAPTER

03

HAZARDS IN CHEMISTRY LAB

INTRODUCTION

Diverse Methods of Exposure to Dangers: This section describes the potential routes of exposure and explains how blocking these routes can prevent exposure. An Introduction to the Safety Industry's Signs, Symbols, and Labels: This book discusses the fundamentals of hazard recognition, a skill required of all laboratory personnel. Such proficiency can be achieved by comprehending labels, signs, symbols, words, and other informational sources. Material Safety Data Sheets (MSDSs) serve as an excellent resource for discovering information about potential hazards posed by chemicals and their usage. [MSDSs] are a beneficial example of a resource that can be used to discover information on potential hazards posed by chemicals, as well as how to use those chemicals. Information and Resources Regarding the Laboratory's Dangers and Safety: This book provides a description of the readily available information resources regarding the dangers of chemicals, as well as instructions on how to use this information to recognize the dangers posed by chemicals. MSDS Information and Its Interpretation: This section provides an overview of the many components that comprise a typical MSDS as well as suggestions for locating relevant information regarding a chemical's hazards.

Acute Toxicity: In chemistry classes, this chapter discusses and provides examples of the numerous categories of acutely hazardous compounds that pose immediate concerns. Chronic exposure to toxins: This book provides a concise introduction to the concept of chronic toxicity and explains why it is essential to comprehend this type of toxicity when working with chemicals over extended periods. Biological Dangers and Safety Precautions: This book explains the hazards posed by biological agents, including where and how they could be encountered, along with some general strategies for preventing exposure. When working in his research laboratory, the user knew it was essential to take precautions to avoid exposure to the chemicals he was handling. He wore chemical goggles and gloves and used a chemical hood during his experiments to protect himself from the potentially hazardous substances. In order to secure his hands while working with the chemicals in the morning, he wore rubber gloves. When he returned home in the afternoon, he promptly donned the gloves and resumed his work. After some time had elapsed, he became aware that his skin had begun to redden and burn. Then he realized that he had placed the gloves on backward, meaning that the chemically contaminated exterior was now in direct contact with his skin. Due to the severity of the exposure, he required medical attention, was unable to work in the laboratory for two months, and was ultimately compelled to abandon this research project after developing an allergy to one of the chemicals.

A potential threat or source of damage is known as a hazard. When we say that something has the potential to be hazardous or harmful, we mean that it is capable of doing so. When we speak about exposure, we mean coming into direct contact with a hazard or chemical in a way that can cause injury or harm. User could acquire a dose of a chemical by ingesting it, inhaling it, or allowing it to come into contact with the user's skin; in other words, the user could acquire the dose by allowing the chemical to penetrate the user's body. user's bocy. User could also receive an additional dose, such as trauma from being struck directly by an object [shrapnel] from an explosion or indirectly by a secondary object that was impacted by another moving object. Both of these possibilities are plausible. Numerous factors, including dose, duration of exposure, and route of exposure, can have a significant impact on the severity of an injury or harm. Remember that our goal is to minimize or eliminate users' exposure to potential hazards. This is the most essential point to remember.

The routes by which chemicals may reach or penetrate the body are known as "routes of exposure." In order for a chemical to have the potential to cause harmful effects in a living organism or body, it must find a way to come into contact with and penetrate it. To safeguard oneself from the effects of harmful chemicals, it is sufficient to prevent them from entering the body or exposing the body to them. If users are aware of the various routes of exposure and take steps to prevent chemicals from accessing users' bodies via those routes, users effectively eliminate their exposure to the chemicals in question and protect themselves from any potential damage they may cause. Four methods of entry or exposure are ingestion, inhalation, contact with the skin and eyes, injection, and topical application.

A hazard is a risk or potential for damage, and there are numerous hazards in the world today.

- Swimming is a lot of fun, but if certain conditions are met, the water can pose a significant risk to people who aren't strong swimmers and even to those who are, if they aren't careful. In reality, water is responsible for the deaths of more than 3,500 individuals annually in the United States. Nevertheless, we cannot endure without water because it is essential to our very existence.
- An uncontrolled fire poses a significant risk, but when we have it under control, it becomes a valuable tool that we can use to keep warm or prepare food.
- Many of us rely heavily on air travel for long-distance travel, but accidents involving aircraft do occur, albeit infrequently; flying is a safer mode of transportation than operating a car.
- Numerous individuals rely significantly on air travel. Automobiles are the primary mode of transportation for the overwhelming majority of United States residents; however, if these vehicles are operated unsafely, they pose a substantial risk of injury or death. 40,000 individuals are killed annually in automobile-related incidents. We take a large number of medications that alleviate our distress, treat our ailments, or protect us from disease; however, if we do not use them properly, they can pose serious risks to our health.
- Numerous items, such as the chemicals we use on a daily basis, offer both substantial benefits and very real risks. This is especially true for numerous other things. The fact that we can learn how to optimize the benefits of these hazardous materials while minimizing the risks associated with them by taking specific precautions is a positive feature of the hazardous materials themselves. [47]

Users are exposed to substances and other materials and procedures in the laboratory, which, like everything else in the world, have both positive and negative consequences, such as the potential for chemical reactions that can lead to accidents or the benefits of conducting experiments that advance scientific knowledge. Essential to ensuring not only users' own safety but also the safety of those around users is the ability to recognize hazards and manage, control, or mitigate the risks posed by them. This section presents the fundamentals of risk recognition in the laboratory, also known as the language of hazards. Typically, the working environment in a laboratory is quite secure. If a user operates in a laboratory, however, the user may be exposed to hazards of a chemical, biological, radiation, and physical nature, which can lead to serious health risks if not properly managed and mitigated.

FINDING HAZARD INFORMATION: MATERIAL SAFETY DATA SHEETS [MSDS]

Three female pupils were using 1-fluoro-2, 4-dinitrobenzene [FDNB]. When they worked with this reagent for between two months and one year, the tips of their fingertips and toenails became stained with the characteristic yellow hue of FDNB. Additionally, their fingertips developed blisters. Testing revealed that both the students and their 39-year-old instructor had developed sensitivity to FDNB. Further research revealed that it was historically recognized as both a potent allergen and a significant irritant. The students hypothesized that their gloves, which allowed FDNB to travel through them, caused their sensitization. After examining their protective garments, it was determined that FDNB could penetrate them. [48]

The Chemical Abstracts Service [CAS] of the American Chemical Society disclosed in September 2009 that it had CAS numbers for 50 million chemical substances and that the list was growing, necessitating a future transition to 10-digit CAS numbers. The American Chemical Society released the report to the public. The increase in the number of distinct chemicals is depicted in the graph below. Information about the risks posed by most of these chemicals is scarce because many chemicals are not evaluated for their potential toxicity, and most of these compounds have not been investigated. In some instances, the available information must be actively sought by searching for specific information pertinent to a particular molecule, and it may still be difficult to locate, especially when databases are incomplete or when the chemical is newly synthesized and lacks comprehensive studies. [49]

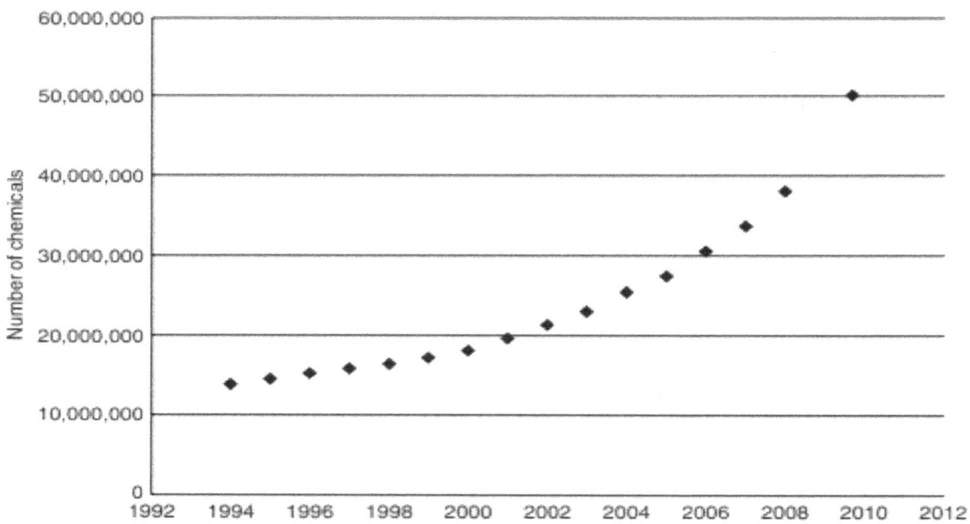

FIGURE 3.1.3.1 Number of CAS Registered Chemicals.

For the much smaller number of chemicals manufactured and sold in the United States, perhaps less than 0.1% of those reported by CAS, we rely on chemical suppliers to provide us with information regarding the known and documented hazards associated with those chemicals. However, chemical suppliers can only report the information they have access to, and frequently, there is only a limited amount of hazard information available; this is notably true in terms of the chemical's toxicological properties. Despite these limitations, users should never assume that exposure to any chemical is risk-free, and users should always make an effort to learn as much as possible about the potential hazards posed by the chemical in question. [50]

The fact that the toxicological effects of more than 99 percent of known chemicals are unknown is, therefore, a disappointing development. The favorable news is that we have at least some toxicological information for the overwhelming majority of substances encountered in an undergraduate chemistry curriculum. Consequently, for the majority of the chemical users who are likely to use during their time as a student, users have the opportunity to learn about their health effects. [51]

The Evolution of Material Safety Data Sheets: From Hieroglyphs to the Internet

Material Safety Data Sheets MSDSs are one possible resource to consult if users wish to learn about the potential hazards associated with a particular chemical. Chemical distributors produce MSDSs as documents. These documents outline the chemical's characteristics and safety-related information. In 1980, the Occupational Safety and Health Administration [OSHA] drafted the Hazard Communication Standard, which mandated that all chemical suppliers provide Material Safety Data Sheets [MSDSs] to chemical consumers. The ancient Egyptians recorded some of the information found on Material Safety Data Sheets [MSDSs] using hieroglyphs. Archaeologists have unearthed these hieroglyphs in tombs and on papyrus documents. 5 Material safety data sheets are readily accessible on a variety of websites today. Numerous chemical suppliers include a hard copy of the Material Safety Data Sheet [MSDS] in the packaging of the chemical they dispatch to any business or individual making their first purchase of that chemical. Alternatively, some organizations may offer to provide users with access to their online MSDS database. [52]

The purpose of creating Material Safety Data Sheets [MSDSs] is to assist users by providing information about the potential hazards posed by a chemical. These documents must contain at least some basic information about the substances they discuss. Material Safety Data Sheets [MSDSs] must contain specific information, and OSHA advises—but does not mandate—that all MSDSs adhere to the American National Standards Institute [ANSI] Z400 format. This format is used to present the information in Table 3.1 for users' perusal. In addition, the newly proposed Globally Harmonized System of Classification and Labeling of Chemicals [GHS, see Section 3.2.1] recommends using the same format for Safety Data Sheets [SDS] worldwide. [53]

Material Safety Data Sheets [MSDSs] are widely available and simple to obtain, but there is no legal requirement that they be accurate. Therefore, while it is reasonable to assume that the data provided is intended to be accurate, the documents typically include a disclaimer [in small type] that absolves the company of responsibility for the information's accuracy. This disclaimer is done so that the company can avoid liability for any possible errors. This practice is one of the reasons why many safety experts do not rely on MSDSs as a reliable source of the most up-to-date and accurate information regarding workplace hazards. [54]

American National Standards Institute [ANSI] Material Safety Data Sheet Format

1. Identification
2. Hazard [s] Identification

3. Composition/ Information on Ingredients
4. First Aid Measure
5. Firefighting Measures
6. Accidental Release Measures
7. Handling and Storage
8. Exposure Controls/ Personal Protective
9. Physical and Chemical Properties
10. Stability and Reactivity
11. Toxicological Information
12. Ecological Information
13. Disposal Considerations
14. Transport Information
15. Regulatory Information
16. Other Information

In the subsequent paragraphs, we will study each section of an MSDS in greater detail. Let's begin by examining the broad picture of these materials. First and foremost, users should consult the MSDS for information on the potential hazards posed by the chemical they are about to use. The user should also research any precautions or special handling instructions for this chemical. As the user may have guessed from a previous statement, not all MSDSs are created equally in terms of quality; some are good, some are not, some may contain incorrect information, and a few may not provide sufficient information for the user to accurately assess the risks associated with the chemical. There may also be inconsistencies in the Material Safety Data Sheets [MSDSs]. For instance, one section of the MSDS may state that ventilation is necessary to keep air concentrations of a chemical below the permissible exposure limit, while another section may state that there is no established exposure limit. This type of inconsistency is typical. Nevertheless, a Material Safety Data Sheet [MSDS] is a wonderful place to begin users' search for information regarding the dangers posed by a chemical. It is recommended that the user investigate the MSDSs provided by other manufacturers for the same chemical to determine if the hazards are stated more clearly. First responders, particularly those working in fire departments and medical emergency services, such as doctors, can use MSDSs as a crucial source of information in emergencies. Numerous businesses maintain physical copies of these documents at all times so that they can be provided to first responders in the event of a spill requiring assistance from outside sources or an exposure injury requiring specialized medical attention. [55]

Finding Hazard Information in MSDSs

Before using a material safety data sheet [MSDS], the user should first verify that the user has the correct MSDS by comparing the information on the product's label to the information supplied at the beginning of the MSDS. There is a link to various MSDS sources in Reference 6. Examining a Material Safety Data Sheet [MSDS] can help users understand the structure of these documents. The subsequent section provides additional information on how to interpret an MSDS. Verify that the name on the bottle matches the name on the bottle's label, and proceed with caution when reading the name, as the names of many substances are exceedingly similar. Compare the CAS number the user discovered to the one the user had previously.

The second stage involves gathering as much information as possible about the potential dangers posed by the chemical in use. User may need to search multiple sections, but it should be near the start of at least one. Observe sections with titles such as "Hazard Identification," "Physical and Chemical Characteristics," "Fire and Explosion Hazard Data," "Reactivity and Stability Data," and "Health Hazard Data." There is a chance that later sections of the MSDS contain additional information that may aid user in determining the level of risk posed by this chemical.

User must search for terms analogous to those discussed in this section, such as corrosive, flammable, toxic, and irritating. For instance, the MSDS for acetone may state, "Danger! Extremely Flammable." Other MSDSs may use a slightly different phrase to describe acetone's hazards, such as 'flammable liquid and vapor.' "Vapor could cause an unexpected conflagration." It could include the ubiquitous phrase "Dangerous if ingested or inhaled."

In other instances, the risk may not be described as exhaustively as it ought to be. Under "potential health effects," users may find a section titled "Inhalation," which explains that inhaling vapors irritates the respiratory tract or that exposure can cause irritation of the respiratory tract and pharynx. It is not difficult to determine that this substance is skin-irritating. On the other hand, it may state that overexposure or excessive concentrations may cause depression of the central nervous system, unconsciousness, respiratory failure, and mortality. These effects are extremely detrimental. In addition, it may contain medical terms that users are unfamiliar with or do not comprehend, such as narcosis, dyspnea, asphyxia, or emphysema. Given that the application of these terms can occasionally lead to perplexity, users may wish to seek assistance to comprehend what they mean. The user can locate an excellent online glossary of safety terms at http://www.ilpi.com/msds/ref/index.html. The website is quite navigable. However, users should be aware that the adverse effects vary from relatively minor to occasionally severe as exposure increases. Keeping these factors in mind is important. [56]

Learning About Safe Handling Precautions

After educating users on the dangers posed by the chemical, users must then educate themselves on the exact protocols for its safe management. As an example, it may state, "Keep away from heat, sparks, flames, and other ignition sources." Typically, MSDSs contain information that is relevant to firefighters. It may imply, for instance, that water is ineffective for extinguishing fires caused by this chemical and that other extinguishing methods should be used instead. One of the sections of this document discusses personal protective equipment [PPE] and exposure controls, including eye protection, skin protection, and respiratory protection. A significant issue is that the Personal Protective Equipment [PPE] section of many MSDSs is far too general to be useful. For instance, the MSDS might recommend wearing protective gloves to protect users' skin from exposure, but it's important to know which gloves are considered suitable. No glove works for all compounds; each substance needs a specific one. There are also protocols that define how to handle emergency situations, such as spills, although the advice in these manuals is frequently too vague to be of much use. For instance, the recommendation to "follow state and local protocols" is accurate, but it provides no assistance. [57]

Reading MSDSs with a Critical Eye

Material Safety Data Sheets [MSDSs] are frequently used in assessing the qualities and risks associated with a specific chemical because they are so readily available. As stated previously, they are not always as reliable or precise as they could be, but they are a useful starting point for gaining a general understanding of the dangers posed by a particular chemical. Depending on the specific application the user has in mind for a chemical, it is highly probable that the user needs to conduct additional research into its properties. Many chemists believe that Material Safety Data Sheets [MSDSs] occasionally exaggerate the dangers posed by a chemical, its potential adverse health effects, and the need to wear protective equipment. It is not difficult to hypothesize that this practice arose as a result of concerns about the potential for legal liability resulting from the use of this substance and the legal counsel that resulted from those concerns. Although this behavior is understandable, it can make it more difficult to identify the most significant hazards associated with a chemical, the most significant adverse health effects, and the necessary personal protective equipment [PPE] in a laboratory setting. Remember that MSDS (Material Safety Data Sheets) were not designed for use in laboratories but rather for the industrial application of the chemical, which is more likely to involve the use of significant quantities of chemicals.

Users learn more about other sources of information [besides MSDSs] in this section. [58]

Using MSDSs in the Lab:

RAMP

RAMP is straightforward to implement when working with MSDSs.

The Material Safety Data Sheet [MSDS] is a valuable source of information that can be used to:

- Identify the hazards associated with a particular chemical.
- Determine the quantity of risk posed by the chemical based on its physical characteristics.
- Prepare for emergency situations based on the assessed hazard level and the anticipated effects of exposure.

INFORMATION RESOURCES ABOUT LABORATORY HAZARDS AND SAFETY

In the laboratory, a technician was preparing some samples using a method that included the use of isopropanol. After he had finished, he poured the remaining isopropanol into a plastic container and labeled it "2-Propanol." The container exploded instantly, spraying its contents all over the laboratory worker, and the reaction occurred very rapidly. Despite being stunned, he still managed to call for help. His coworkers rushed to his aid when he complained that his skin was burning. After he explained the situation, they used an emergency eyewash and an emergency shower to remove the splash. After receiving treatment for his acid burns, he was evaluated and ultimately discharged from the university medical center to which he was transferred for care. The investigation revealed that the bottle's label was a forgery and that the contents were actually concentrated nitric acid and copper waste. [59]

What Specifically Does the Globally Harmonized System [GHS] for the Classification and Labeling of Chemicals Encompass?

Due to the global nature of the modern marketplace, we import food, products, and raw materials from all over the world and use them in our production. A group of United Nations-affiliated nations has reached an agreement to create a classification and labeling system for hazardous chemicals. The objective of this initiative is to ensure that all nations classify and label chemicals in the same way so as to enhance international interactions within the chemical industry. The acronym GHS refers to a program that may standardize the classification and labeling of chemical substances. In 1992, the United Nations Conference on Environment and Development in Rio de Janeiro, Brazil, authorized a mandate to build the Globally Harmonized System [GHS], and by the year 2000, the system's foundation was in place. The process of developing a global system to classify and designate chemicals has been slow and challenging, and the implementation of this system is still in progress. The initial version of GHS was published in 2003. The Purple Book is the name of the document that describes the GHS; it has been revised three times: in 2005, 2007, and 2009. [60]

The Globally Harmonized System [GHS] is an international system that includes documents comparable to Material Safety Data Sheets [MSDSs], a standard labeling scheme for chemicals, and procedures for determining the hazard ratings of substances. Individuals familiar with Material Safety Data Sheets [MSDSs] in the United States may recognize many aspects of this classification system; however, they will also observe certain modifications because it has been refined and accepted as an international system to enhance collaboration in global trade. Participation in the GHS is completely optional. Using the Globally Harmonized System [GHS] enables countries to sell and purchase chemicals across borders with the assurance that hazardous chemicals are designated and labeled in the same manner as in other GHS countries. This allows transportation and shipping companies to safely manage these chemicals. The GHS makes it possible for countries to implement its system. Common chemical pictograms, signal words, and hazard statements are included on labels to inform the chemical's handler, emergency personnel, and ultimately the chemical's user of the chemical's relative toxicity. These labels inform the user of any unique risks associated with the chemical. To arrive at these classifications through consensus, the GHS had to be established, which was a complex undertaking that required the participation of a large number of nations. [61]

Organization of the Globally Harmonized System of Risk Assessment

The three primary types of hazards are physical dangers, health hazards [for humans], and environmental hazards [focusing on aquatic systems]. The class's distinctive relative ranking system depends on the representation of each category, which may be letters, numbers, or other types of descriptors. In every instance, the letter "A," or the lowest numerical rating, denotes the classes deemed to be the most hazardous, while the classes deemed to be the least hazardous have higher numbers and subsequent letters. Users are not required to read the entire document; rather, it is provided to illustrate the diversity of hazards covered by GHS. For the purpose of identifying hazards, GHS-compliant labeling can be used. For instance, a user may recognize that a low number or an early alphabetic character indicates that a substance could be extremely hazardous. The user will discover in the following sections that the GHS can also be used as a risk assessment instrument. The only difference between the flammability rating system and the NFPA rating system for flammable liquids is that the numerical values in the two rating systems are reversed! The NFPA system allocates "0" to the least flammable materials and "4" to the most flammable materials. Due to this terrible lack of uniformity, it is difficult for chemists and firefighters in the United States to perform their duties because the NFPA system is so extensively used, leading to potential confusion and misinterpretation of flammability hazards in emergency situations. This section presents the GHS flammability ratings for consideration. This section describes the GHS ratings that pertain to prospective health risks. Throughout the entirety of this book, the GHS hazards are discussed in detail. [62]

Implementation

Originally, 2008 was intended to mark the commencement of GHS adoption worldwide. The United Nations maintains an up-to-date website that details the implementation progress of 67 distinct countries. It is possible that GHS 4 adoption was not completed until 2015.

The Occupational Safety and Health Administration [OSHA], the Environmental Protection Agency, the Department of Transportation, and the Consumer Product Safety Commission are the four federal agencies involved in implementation in the United States; however, OSHA has the most influence on laboratories, particularly in enforcing safety standards and regulations that affect laboratory practices and procedures. The Occupational Safety and Health Administration published a proposed rulemaking notice in October 2009. [63]

Using GHS

The Globally Harmonized System [GHS] is a tool that can assist laboratory personnel in determining the level of peril posed by working with hazardous substances. The GHS specifies the format for documents that are similar to MSDS and require the inclusion of the information enumerated below. These documents are known as Safety Data Sheets [SDSs] and contain the following sections of information:

1. Identification
2. The identification of potential threats
3. Composition and information regarding constituents
4. Fundamental life support techniques
5. Methods for extinguishing flames
6. Accident release mechanisms
7. Handling and storage
8. Exposure control/personal protection
9. Characteristics of the physical and chemical constitution
10. Consistency in addition to sensitivity
11. Details pertinent to toxicology
12. Ecological information
13. Disposal consideration
14. Information regarding conveyance
15. Specifics regarding regulations
16. Additional information

They should be perused similarly to Material Safety Data Sheets.

Moreover, the format for the labeling of substances is standardized. On each label, there is a product identifier, a pictogram [symbol], a signal word, a warning about potential dangers, and information about the manufacturer. If the manufacturer or distributor of the chemical so chooses, additional information may be included on the label. The SDS contains a letter or number that corresponds to the category designation listed. This designation exists. Alternatively, if a transport pictogram is present on the packaging, the GHS pictogram for certain hazards may be absent. There are a total of nine pictograms and two signal words: "danger" and "warning." Examples of potential hazard statements include "May cause respiratory irritation," "Toxic upon contact with skin," and "Heating may result in a fire or explosion." In Figure 3, all of the pictograms and the hazards that each one represents are displayed. [64]

FIGURE 3: GHS logos. Once the GHS is approved and implemented in the United States within the next ten years, these pictograms are likely to be extensively used. [With the permission of the United Nations Economic Commission for Europe, 2007 United Nations, New York, and Geneva Copyright.]

RAMP

RAMP benefits from the availability of GHS Safety Data Sheets and GHS labeling.

- User must be aware of the risks posed by a particular chemical.
- Determine the level of danger posed by the chemical by analyzing its physical properties and drawing conclusions based on these analyses.
- Reduce the risk of injury by adhering to the SDS recommendations for the use of personal protective equipment and any other safety equipment that may be required.

- Develop contingency plans based on the evaluated hazard level and potential consequences of exposure.

INTRODUCTION TO TOXICOLOGY

A person conducted the "new" laboratory experiment known as the "Dipoles Experiment." The student created a substance resembling an industrial whitener in three distinct phases over the course of one week. The face, thorax, arms, and thighs of the student all had severe dermatitis at the same time. Participation in the same experiment resulted in the manifestation of varying degrees of dermatitis on the bodies of six additional students. One of the compounds used in the final phase of the experiment was CBPH, which stands for chlorobenzaldehyde phenylhydrazone. Five of these seven students had a positive reaction to the chemical, according to a skin test.

Prelude

The following section provides an overview of the fundamental scientific principles and terminology used in toxicology. In this section, we discuss acute toxicants, which are also known as hazardous risks. These are subjects with which users should become familiar early in their academic career. This section focuses primarily on chronic toxins when discussing carcinogens. In research facilities such as academic laboratories, carcinogens pose a more imminent threat than chronic toxicants. After discussing the biotransformation of hazardous substances in Chapter 3 and the biological hazards found in chemical laboratories in this section, this section of toxicology concludes.

What is toxicology, and why do I need to become educated on it as soon as possible?

The discipline of study known as toxicology investigates the potentially harmful effects of chemical exposure on living organisms. According to the definitions supplied by multiple dictionaries, the term "toxicology" can also refer to the study of poisoning. Toxicology involves much more than the study of toxicity. The vast majority of substances are not considered poisonous [for more information, see the section that follows this one]. Toxicology refers to the capacity of a chemical to induce systemic changes or harm in living organisms. Such an effect could be accomplished by eliminating organs or by altering a biochemical system or process. Toxicology is an area of study within pharmacology. A "systemic" effect occurs when a hazardous material is absorbed into the bloodstream and transported to other body parts. To recognize the potentially hazardous properties of chemicals, it is necessary to have a fundamental comprehension of toxicology as well as the related terminology. Consequently, users are able to comprehend the impact that dose rate has on the potential toxicity of each chemical.

"Toxicant" is the term used to characterize a chemical that is capable of producing toxic effects; the term "toxin" is more appropriate for naturally occurring dangerous substances. Throughout the book, both terms are used interchangeably. People frequently erroneously apply the term "toxin" to all potentially hazardous compounds, which is incorrect. "Toxin" only refers to a subset of naturally occurring toxicants, such as botulinum toxin, which is among the most toxic substances known to humans. "Toxicant" refers to all hazardous compounds, whereas "toxin" refers only to a subset of toxicants derived from natural sources. Toxins can be divided into two groups: organic and inorganic.

Obtaining the Appropriate Dose Is crucial. What brought the user into contact with the chemical in the first place?

When a toxic substance enters or comes into contact with the body, it may exert its detrimental effects immediately. The duration of a person's exposure to a chemical is one of the most important determinants of the chemical's toxicity. This quantity is referred to as the "dose rate." A sixteenth-century physician named Paracelsus is credited with making an observation at the outset of this section. The dose of a chemical, or the quantity that an organism consumes of the chemical, determines whether or not the chemical has negative effects. When determining whether a chemical is harmful, one of the most essential factors to consider is the amount of time that has passed since the dose was administered.

A chemical's acute toxicity refers to its ability to induce systemic damage to an organ or organs, or to a biological system or process, with only a single dose. Even if this is not always the case, the result is typically instantaneous and has immediate consequences. This is not always the case, however. In cases of acute toxicity, seconds, minutes, hours, and sometimes even days are commonly evaluated as time intervals. Compounds that pose an imminent threat to human health are provided in the following paragraph.

Chronic toxicity is the capacity of a chemical to cause systemic damage to an organ or organs or a biological system or process with recurrent lower doses over an extended period of time, typically months or years. Chronic toxicity can affect numerous organisms, including humans. Chronic toxicity can result from recurrent exposure to a drug in progressively smaller doses. The dose causing acute toxicity is significantly higher than the dose causing chronic toxicity. Chronic toxicity can be caused by receiving modest doses continuously or multiple doses over an extended period of time. This dose is significantly lower than the acutely hazardous dose. If the body stores small quantities of a chemical until it accumulates enough to cause damage, the chemical can be considered an accumulative poison despite its chronic toxicity. This may occur if the substance in question has an adverse effect on living organisms over time. In the table located in this section, the reader is provided with examples of drugs that display chronic toxicity.

This distinction is crucial because the effects of acute toxicity are not accurate indicators of chronic toxicity. This suggests that the effects of acute toxicity and chronic toxicity may differ, as they may affect distinct organ systems or physiological processes.

The metabolic process, which is the body's primary defensive mechanism, neutralizes toxic compounds.

Through a process known as metabolism, the user's body is capable of metabolizing potentially harmful compounds. This is an advantageous attribute. The metabolic process can convert toxins into innocuous substances, less toxic, or even more toxic than they were before they were expelled from the body. Such reactions can occur at any point along the path of the contaminant through the body. The metabolic process is a dynamic one that can occur in a variety of ways. If users consume small quantities of a variety of chemicals, it is likely that they are metabolized and either utilized in the body for some function or eliminated from the body before they can pose a significant risk. This is the case if they are metabolized and utilized for some function in the body. Since "dynamic" implies that it is always occurring, it is likely that it occurs if the user consumes moderate amounts of a variety of compounds. This is because "dynamic" implies that something is constantly occurring. Due to this, it is possible to ingest potentially hazardous substances such as ethyl alcohol in beverages or paracetamol in pharmaceuticals while experiencing only the acute effects of these substances for a brief period of time. In a similar manner, we might be exposed to minute quantities of dangerous compounds in the food or water that we consume; however, because the levels are so low, the body's metabolic system is able to eliminate these toxins without negative effects.

If this were not the case, the numerous contaminants that enter our bodies every day would almost certainly harm our health. Damage to the liver, the primary metabolic organ, can impair a person's ability to metabolize pharmaceuticals and other substances, including chemicals. This effect is because the liver is the primary organ responsible for metabolism. The metabolic pathway that toxic substances follow is dissected into its component elements and discussed in greater depth in the following paragraphs. Because certain substances are resistant to the metabolic processes that occur in the body, they are more likely to be stored or accumulated in the body than to be eliminated, which can lead to potential health risks and long-term exposure effects, such as those seen with chlorinated compounds like DDT. In recent decades, the pervasive use of chlorinated compounds has left many of us with detectable levels of these chemicals. DDT is an example of a pesticide with long-lasting effects. Even after several years, these compounds are still present in our environment, and when we ingest them, our bodies store them in fatty tissue. Although higher concentrations of these substances may have deleterious effects, it is widely believed that the quantities of these substances that are stored in our fat are, for the most part, safe.

The process the user is observing is known as "bioaccumulation." [However, not all scientists agree; for more information, see the Special Topic Controversy in Science—Epidemiological Challenges and Degrees of Certainty.]

There is a possibility that when chemicals are metabolized, they become more hazardous than they were originally. Antifreeze's primary ingredient, ethylene glycol, is an excellent illustration of this concept. When ethylene glycol is consumed, it is metabolized into molecules with adverse effects, i.e., it becomes harmless. Among these effects are metabolic acidosis, a condition caused by an excess of acid in the body, and kidney and brain damage caused by the formation of calcium oxalate crystals. Eventually, ethylene glycol poisoning can result in mortality.

Toxicologists discuss lethal concentrations, which they refer to as LD50 and LC50 values, respectively.

Over a number of years, scientists and medical professionals have investigated reports of accidental poisonings, homicides, suicides, and occupational exposures to humans to gather information regarding the detrimental effects of chemicals. Despite these efforts, the experience and data available to us are relatively limited, making the dosage predictions susceptible to some limitations. Studies on animals are the most informative category of toxicology research; all other types of toxicology research provide less information on toxicity. To evaluate the toxicity of potential chemicals, toxicologists typically employ rats, mice, and other types of laboratory animals to conduct the necessary tests. As a result of their investigation, they have compiled a list of key terms applicable to the process of determining a substance's degree of toxicity. When discussing the acute toxicity of a substance, the term "lethal dose fifty" [LD50] is commonly employed. The value is commonly measured in milligrams [mg] of the chemical per kilogram [kg] of the animal. When dosages are expressed using units such as mg/kg, it is substantially easier to evaluate the differences in concentrations required for various species. The solitary dose, known as the LD50, is the amount of a substance that results in the death of fifty percent of the test animals. This dosage has been extrapolated or calculated rather than being derived from a measurement taken under actual conditions. Throughout the course of the tests, multiple dosages are administered, and at each dosage, some animals are able to survive while others are slain. For instance, when a specific dose is administered to animals, ninety percent survive, but when a different dose is administered, only seventy percent, twenty-five percent, or ten percent survive. Following this, a graph of the test data points is used to estimate the LD50 value. Studies that include oral or dermal dosing are the ones that utilize the LD50 the most frequently. The graph in the table is generic; the slopes of the lines indicate various compounds, some with extremely steep slopes and others with more moderate slopes.

When assessing the toxicity of a substance through inhalation, the term LC50, which stands for "lethal concentration fifty," is employed. The median lethal dose, or LD50, is the lowest dose that causes mortality in 10% or more of the tested animals. This is an extrapolated dose, expressed as the air concentration of the potentially hazardous chemical. Typically, it is expressed as parts per million, milligrams per cubic meter, or micrograms per liter. [Please refer to the section for additional information on ppm.] The cost of determining LC50 values is high because it requires additional equipment [inhalation chambers and detailed measurements of air concentrations] that is not required when assessing lethal doses via oral administration, skin administration, or any other route. The majority of currently accessible experimental toxicological data is presented as LD50 values. This disparity is due to the fact that determining LD50 values is considerably easier and less expensive than determining LC50 values. In toxicity research, rats and mice are frequently used as test subjects. This is due to the fact that smaller animals are simpler and less expensive to care for.

Extremely Dangerous to Human Substances, Also Known as Drugs

What is the precise definition of poison? Existing non-hazardous substances leave considerable room for interpretation when it comes to the meaning of the word "poison," particularly in the media, literature, and even among scientists. Due to this difference in usage, there has been some confusion regarding the meaning of the word "toxic." This confusion may be because we frequently use the terms "poisoned" and "poisoning" to describe situations in which a person intentionally or unintentionally consumed a chemical dose that caused them to become gravely ill or even die. In contrast, poisoning refers to the act of causing another person to become gravely ill or die. However, even if it is reported that a chemical "poisoned" a person, the term does not necessarily mean that the substance in question was a poison.

A chemical is considered toxic if, even in minute quantities and after ingestion or exposure, it can cause severe illness or mortality. This category includes toxin-containing substances. When discussing what constitutes a poison in this volume, we employ toxicologists' standard terminology, such as LD50 and LC50 values. The chemical is toxic if its oral lethal dose [LD50] is less than 50 mg/kg and its dermal lethal dose [LD50] is less than 200 mg/kg. Poisonous substances are frequently referred to as "highly toxic."

The inhalation LC50 is less than 200 ppm or 2 mg/L [dust], whichever value is less. Toxic compounds are always present in poisons, but not all hazardous chemicals can be classified as poisons. Poisons are always composed of toxic substances. This doesn't mean the user can't be poisoned by other chemicals; it means the user can be poisoned by non-poisons, suffer severe symptoms, or die from exposure to these chemicals if the dose is high enough and the exposure is long enough. On the other hand, due to their high level of toxicity, poisons pose a substantial threat to anyone who comes into contact with them. Due to their toxicity, they are frequently marked with "Poison," "Danger," and the standard "Skull and Crossbones" symbol. There are only a few categories of substances that can be classified as toxic.

There is a widespread perception in today's modern society that it is possible to generalize about the toxicity of a drug based on whether or not it is a naturally occurring or synthetic [man-made] compound. This is a fallacious supposition. Contrary to this assumption, this is not the case. It is essential to remember that neither "naturally occurring" nor "synthetic" should be interpreted as indicating a particular level of relative safety or peril. In actuality, the substances commonly recognized as being the most hazardous are also naturally occurring. Consider botulinum toxin, which is reported to have a lethal dose for humans of approximately 0.000001 mg/kg. Clostridium botulinum bacteria are responsible for the production of botulinum toxin, but the toxin can also be produced accidentally when certain alkaline or low-acid foods are improperly prepared and preserved. Poisoning is a potential outcome, with a fatality rate ranging from 5 to 10% in documented cases of the condition. This neurotransmitter serves as the product's primary active ingredient in the cosmetic treatment known as Botox.

The degree of toxicity is influenced by a variety of factors.

In addition to the amount ingested, a chemical's level of toxicity can also be affected by a large number of other factors. The figure illustrates their most significant characteristics.

It is conceivable that the manner in which a person or animal is exposed to a harmful chemical could have a significant impact on the chemical's effects. When conducting experimental toxicity studies, oral administration is frequently employed in comparative studies. Gavage is a method for administering dosages to animals in which a toxic chemical or its diluted form is administered directly into the animal's stomach via a tube. This route of administration is known as intragastric administration. Oral doses can be administered by consuming the toxic substance while it is mixed with food or water or by gavage. The process is depicted in the figure, which illustrates how the toxicity may vary based on the pathway. It is possible to deliver a toxic chemical to the skin via the dermal route using a patch or paint containing the chemical. Both of these approaches are viable alternatives. Inhalation is another method that entails exposing an animal to a chemical at a specified concentration and for a specified period of time through the air. This procedure takes place over a specific time frame. The term "passive exposure" is used to characterize this scenario.

The figure demonstrates that, when it comes to the administration of toxic doses, the oral route is more effective than the cutaneous route in the majority of cases. It is significantly more difficult to compare the inhalation route with other methods, such as the oral or cutaneous route, due to the wide variety of units, which complicates the assessment of exposure levels and their effects on health. Additional information is necessary to calculate the amount of the chemical that an animal inhaled. This data must specify how much of the airborne concentration the animal ingested during the specified time period. The amount varies based on the size of the animal, its physiology, and the activities it engages in. It is difficult to ascertain how the dose received via inhalation compares to the dose received via other delivery methods, such as ingestion or dermal exposure, which can also significantly affect the overall toxicity experienced by the animal. One of the most crucial aspects of toxicology is the variety of organisms that the contaminant harms. As stated previously, there is no data regarding the toxicity to humans; therefore, the toxicity to animals is used as a general guide. This is the case despite the fact that human safety is almost always our top priority when it comes to toxicity issues.

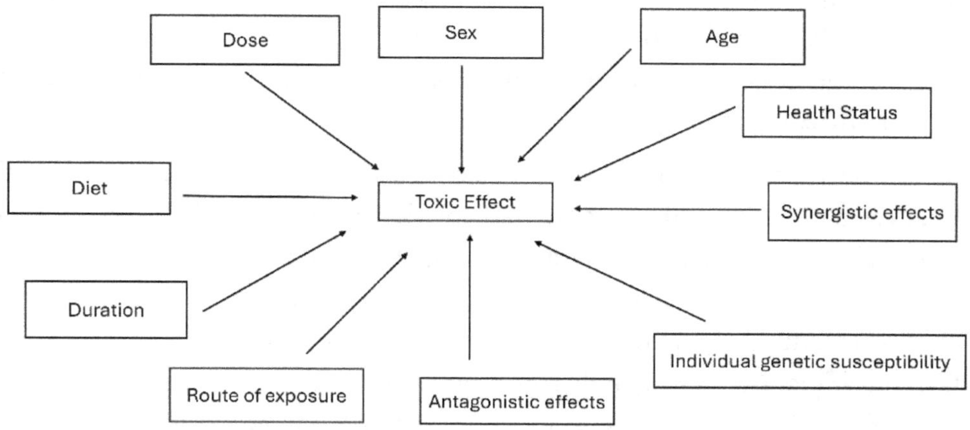

FIGURE: Factors Affecting Toxicity. Many factors determine the toxic effect of a substance for a particular individual. It is not surprising that not all people exhibit the same effect from a particular exposure.

Substance	Species	Oral LD
TCDD	Hamster	1157μg/kg
TCDD	Mouse	114μg/kg
TCDD	Monkey	2μg/kg
TCDD	Guinea pig	0.5μg/kg

TABLE: Variability of Toxicity by Special

However, the effects of potentially hazardous substances on humans can only be approximated, and in some cases only approximated, via animal experimentation. Humans are significantly more complex than animals. Different animal species, such as mice, rats, guinea pigs, rabbits, and primates, may experience substantially different LD50 doses and toxic effects [see Table]. Chemicals may have an effect on only one or two species, but they may also have an effect on others. The data presented in the table indicate that the guinea pig is more than 2,000 times more sensitive to the effects of TCDD than the hamster. Following its identification as the dioxin found at numerous contamination sites in Seveso, Italy (1976), Love Canal, New York (1979), and Times Beach, Missouri (1985), TCDD, also known as tetrachlorodibenzo-p-dioxin, gained widespread notoriety. Using the same poison, an assassination attempt on the life of Ukrainian President Viktor Yuschenko occurred in 2004. However, the scheme was thwarted.

There is no single animal that can accurately represent all of the potential dangers that humans confront. To the extent that it is feasible, toxicologists take into account data obtained from experiments conducted on numerous animal species. Despite this, it is still possible for humans to experience adverse chemical reactions that differ from those observed in animals. This is notably true for effects like cancer that manifest over an extended period of time. Despite this, decades of research have demonstrated that testing for toxicity in animals can frequently and accurately predict the effects on humans. If this were not the case, it would have been unthinkable for us to develop the currently available treatments. We are able to provide accurate estimates of the effects of chemicals on humans by relying on the expert scientific opinion of toxicologists and pharmacologists.

Age is one of the potential components in toxicology. When children are born, their bodies are not completely developed, making them more susceptible to the harmful effects of toxins than adults. Extremely young or elderly individuals are frequently at greater risk of being adversely affected by the toxic effects of chemicals. It is possible that as mature defense systems age, they begin to exhibit signs of frailty or vulnerability. Consideration of the person's current state of health is also pertinent to this discussion. Diabetes, cardiovascular disease, cancer, and arthritis are among the chronic diseases and health issues that are more likely to affect older individuals. The overall state of our health is deteriorating and becoming compromised due to these factors. As a result, we are at an increased risk of being injured by the chemical or biological agents present in our environment. The nutrition of the animals affected the toxicity of the substances to which they were exposed, as revealed by animal research.

The most significant cause for concern is the possibility of negative effects of sexual activity on the reproductive process. Although a few facts about this topic are already known, there has not been a great deal of research conducted on it, with the exception of studies on various birth control methods. In contrast, it is now common knowledge that exposure to certain substances can affect a person's reproductive health. For example, males who are occupationally exposed to dibromochloropropane are more likely to become sterile because of their exposure.

It is possible for a person's genetic makeup to affect how susceptible they are to certain substances, but such an effect is rarely the case. The question of whether or not certain adverse events are truly attributable to our genetic makeup as opposed to detrimental environmental exposure has been the subject of much debate for many years. And as it turns out, distinguishing between these two perspectives is exceedingly difficult because both are viable causes of undesirable events. This long-standing theory suggests that those who are "genetically susceptible" may be more likely to experience adverse effects from toxins. Regarding cancer, this statement is notably accurate. We have learned over a number of years that heredity plays a significant role not only in our health condition but also in each individual's susceptibility to toxic substances. The fact that heredity is an inherited trait facilitated this realization. These two discoveries have completely transformed the course of human history.

The vast majority of published research on toxicants begins with the premise that each substance operates independently. This is not always the case, as numerous exceptions exist. We contend that a synergistic effect occurs when the consumption of a mixture of two substances results in an effect that is more harmful than the "addition" of the effects of each substance consumed individually. The main point of this discussion is that the whole is greater than the sum of its parts. There is also the prospect of an antagonistic effect, which occurs when the total harmful effect of a mixture of hazardous substances is less than the sum of the harmful effects of the individual substances. This can occur when two or more toxic substances combine in a reaction. Medical practitioners and pharmacologists who prescribe or administer drugs must have a comprehensive understanding of these two effects.

Utilizing potentially hazardous materials

Understanding the toxicity of the chemicals used in laboratories is of the utmost importance. This is the case not only due to the variety of toxic substances and routes of exposure but also due to the presence of both immediate and delayed effects, which can lead to serious health consequences if not addressed promptly. In the event that a toxic exposure has occurred, whether it is known or suspected, it is strongly recommended to contact the lab instructor or supervisor as soon as possible to receive guidance on how to manage the situation most effectively.

RAMP

When working with substances that can be inhaled, it is important to wear protective gear such as gloves and chemical hoods. Be cautious around potentially dangerous substances. Determine the extent of danger based on the probability of exposure via each route. Reduce the risk of injury by adhering to safe laboratory procedures and wearing protective gear.

In the event of an emergency, users must be aware of the potentially hazardous effects that the compounds they are handling could have, such as toxicity, flammability, or reactivity, which could pose serious risks to users' health and safety.

ACUTE TOXICITY

They were working in a laboratory, and nearby was a small storage area where they took their lunch breaks. Five technicians were employed there. Three of them reported feeling uncomfortable after consuming multiple cups of tea, with their heads feeling "fuzzy" and their pulses pounding rapidly. Additionally, they had an accelerated breathing rate. Another employee drank two glasses and shortly thereafter began experiencing chest pains that resembled cramping. When he was transported to the hospital, he was treated in accordance with the presumption that he was suffering from a heart condition. Even after consuming only half a cup of tea, a second technician began to feel uncomfortable. One of the technicians recalled that she had substituted laboratory-distilled water treated with sodium azide for the tea kettle to inhibit the proliferation of microorganisms. This is a common technique used in many laboratories. Each and every technician made a complete recovery.

A vast array of toxic chemicals can result in negative outcomes.

In this section, we are studying extremely hazardous substances. These are substances that, when consumed in sufficient quantities, are capable of producing immediate adverse effects. Because their immediate effects are easier to identify and investigate, we have a greater understanding of the effects of acutely toxic substances. On the other hand, it is likely that we do not have a comprehensive understanding of the detrimental effects of chemicals, especially those with chronic or long-term effects. In Section 4.1.1, the user learned that the effects of compounds depend on a variety of circumstances. Among these factors are the dose and duration of exposure. Despite Hawthorne's description of salt as a "holy" substance, children have been known to die as a direct consequence of salt poisoning. Salt has a half-life of 3,000 milligrams per kilogram. Although our bodies can function properly with a negligible amount of salt, we have a tendency to add excessive levels of salt to the food we eat. Caffeine is poisonous and has an oral LD50 in rodents of 192 mg/kg; however, a user would need to consume close to 100 cups of coffee in a brief period of time to achieve a lethal dose. The section states that "the dose makes the poison." There are chemicals that are known to have numerous negative effects, and the processes underlying these effects may vary considerably.

In contrast to the millions of known compounds, however, our understanding of the consequences of toxicity is relatively limited. Our understanding of acute toxicity is based on what we have learned over the centuries. Only a few thousand compounds out of the millions that exist have been the subject of research. It is not uncommon for scientists to make a startling discovery about previously unknown side effects of a medication. This is due to the fact that the subject at issue has never been studied before. In the middle of the 1970s, it was discovered that vinyl chloride [Figure], a chemical used extensively in the manufacturing of polyvinyl chloride [PVC] polymer, caused an uncommon form of cancer called angiosarcoma. This chemical had been extensively used. On occasion, potentially harmful compounds are discovered by accident, and on other occasions, they are discovered through meticulous testing. As a result of the investigation, a substance that was previously believed to be "safe" in the absence of contrary evidence has been determined to be toxic in a manner that was not previously acknowledged. This result should not be surprising to anyone. Recently, it was discovered that diacetyl, a component of the butter flavoring in microwave popcorn, can cause bronchiolitis obliterans, an uncommon form of severe lung disease with the potential to be fatal.

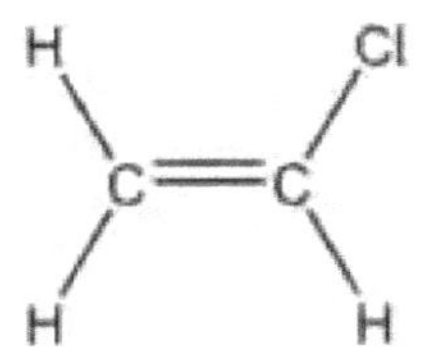

FIGURE: Structure of Vinyl Chloride. The EPA has classified vinyl chloride as a Group A human carcinogen.

We must avoid chemophobia, an irrational fear of chemicals, because everything in the universe, including us, is made of them. Even though some chemicals have the potential to be harmful, we must avoid developing a phobia of them. It is essential to take precautions to minimize or reduce exposure to the various chemical components when working with chemicals. This should be done whenever feasible. Over many years, we have perfected the processes and protocols required for the safe management of a wide variety of potentially hazardous substances, allowing us to do so in laboratories today. If users adhere to these precautionary measures, even if they do not know all of the potential adverse effects of a chemical, they significantly reduce their likelihood of experiencing adverse effects from chemical exposures. This statement is true even if users are unaware of all the potential adverse effects of a chemical.

The fact that users are extraordinarily sensitive occasionally irritates me.

Irritants are substances that do not corrode the skin but cause inflammation at the site of contact. This inflammation could manifest as erythema, thickening, dermatitis, scaling, or blisters. They have the potential to irritate the epidermis, and the results are in every way unsatisfactory, especially from an aesthetic standpoint. Contact dermatitis is the most prevalent occupational skin disorder in the United States. This condition is primarily due to the widespread presence of environmental irritants. According to the test results, about one-third of the 1200 different compounds were irritants. The most frequent adverse effect of toxic substances is irritation. This indicates that there is a substantial danger of coming into contact with an irritating substance while working in a laboratory.

Clinically, there is no distinction between irritant contact dermatitis and allergic contact dermatitis, the two forms of contact dermatitis. When an irritant comes into direct contact with the epidermis, the epidermis reacts in a non-immune manner. The cause of allergic contact dermatitis is a delayed immune-mediated hypersensitivity reaction, which is closely related to but distinct from atopic dermatitis. The section that follows this one contains a list of frequent allergens. Numerous chemicals that can cause cutaneous irritation are detailed. The following are examples of these compounds: Even in low concentrations, corrosives can cause irritation, but contact with highly corrosive or reactive compounds can result in chemical burns and tissue loss.

Chemicals are referred to as sensitizers or allergens when the mechanism through which they exert their effects is the induction of an abnormal immune response. The term "hypersensitivity" is used to characterize this abnormal response. When sensitizers and allergens are combined with acute toxicants, it is generally accepted that there is a latency of a few weeks between the initial exposure and the subsequent hypersensitive reaction. This is due to the rapid toxicity of acute toxins. The firm relationship between these two factors explains why this is the case. Despite this, we classify these substances as acutely toxic due to the frequently severe acute response to the toxin following sensitization. This reaction is typical of acutely toxic substances. The hypersensitive reaction is only possible if the initial exposure leads to sensitization and subsequent exposure to the allergen in question.

Small molecules, such as those found in certain laboratory chemicals, have the potential to function as haptens, thereby contributing to the development of sensitivity. Haptens, which are molecules that react with larger proteins in the body, can produce antigens. If the body recognizes these antigens as "foreign" antigens, which is possible, antibodies are produced to combat them. After being re-exposed to this antigen, these antibodies rapidly form, activating their natural response that protects the body from further injury. As a result of the interaction between the antibodies and the antigen, undesirable side effects such as edema, redness, and irritation may occur. After a person has become sensitized to a chemical, even minute amounts of that chemical can induce an allergic reaction. An allergic reaction is characterized by allergic contact dermatitis, hives, a runny nose, sneezing, enlarged, red eyes, and a headache. A respiratory sensitizer has the potential to trigger asthma, breathing difficulties, and, in extremely uncommon cases, anaphylactic shock in those who are already susceptible to developing these conditions. In the absence of immediate treatment, anaphylactic shock is a rapidly progressing condition in which a person's respiratory system becomes severely constricted, ultimately leading to mortality. Genes play a role in determining a person's level of sensitivity. Individuals who are exposed to an allergen do not necessarily develop a sensitivity to that allergen. Why some people are sensitive to a chemical while others are unaffected by it in their bodies is not completely understood.

Users are probably most familiar with the allergies induced by poison ivy, poison oak, and sumac due to their well-known reputations or users' own personal experiences with them. Urushiol is the name of the chemical responsible for the sensitization response. On the other hand, allergic reactions to compounds discovered in a laboratory are not uncommon. The development of a chemical sensitivity can be extremely debilitating, and it may be necessary to make significant lifestyle changes in order to avoid further contact with the sensitizer.

Without oxygen, the human body cannot function correctly.

Asphyxiants are substances that inhibit the body's ability to absorb oxygen or transport it to vital organs. Depriving cells and organs of oxygen results in the mortality of both the individual as a whole and the cells and organs individually. Based on the method by which they remove oxygen from the ambient air, asphyxiants can be divided into two categories: simple and complex. Helium, carbon dioxide, nitrogen, and methane are all considered examples of inert gases. There is no chemical reaction between these gases and the biological tissue with which they come into contact. There is a correlation between the severity of the hazard and the density of the gas. Asphyxiants with a lower density than air, such as helium, "pool" closer to a room's ceiling because helium has a lower density than air. Carbon dioxide and other heavier-than-air molecules have the potential to "pool" near the ground or in low-lying areas. Those who are required to perform their duties in confined spaces must be more cognizant of the risks present, such as the potential for asphyxiation due to the accumulation of gases like nitrogen or helium, which can displace oxygen and create hazardous conditions.

Nitrogen gas is an effective asphyxiant if its concentration in the air is high enough to "dilute" oxygen to a hazardous level. Approximately 79 percent of the air we breathe consists of nitrogen gas. Nitrogen is typically stored in containers in laboratories. A nitrogen gas tank with a leak, which is generally secure, could be hazardous in a confined space if it is too small. When there is less oxygen in the air, the effects on human health are enumerated. Accidental fatalities have occurred in nursing homes due to the tragic connection of a liquid nitrogen tank to the oxygen system that distributes supplemental oxygen to individual rooms. This incident was due to the striking design similarity between liquid oxygen and liquid nitrogen containers. Because of this potential hazard, regulators for nitrogen and oxygen are constructed with what are known as "opposite threads" to reduce the likelihood of making this type of error. There are two types of errors that can contribute to occurrences such as this: failing to read the tank's label and using the incorrect regulator. Both of these types of errors can have severe repercussions.

Asphyxiants are available in various forms, one of which is the chemical asphyxiant, which inhibits the body's ability to absorb oxygen through chemical means. Carbon monoxide is the most common molecule that can cause asphyxiation and is the leading cause of poisoning in the United States. It kills more than 500 people annually and hospitalizes over 15,000. Carbon monoxide is a byproduct of combustion and can occur in various residential and commercial appliances, such as gas furnaces, water heaters, and open-flame sources such as propane barbecues. Carbon monoxide is a potentially lethal chemical capable of causing brain damage and mortality. This type of exposure is possible in both residential and commercial settings. Carbon monoxide is a colorless and odorless gas with an LDLO of 4000 parts per million [ppm] or 4570 milligrams per cubic meter [30 minutes] for humans. In addition, it exhibits a level of toxicity that is considered moderate. Between 500 and 1000 parts per million [ppm], symptoms such as headache, dizziness, nausea, and fatigue begin to manifest. These symptoms are frequently misdiagnosed as influenza. These are the most prevalent sources; intentional or accidental poisoning accounts for the vast majority of LDLO levels.

On a molecular level, its toxicity is readily explicable, and it is believed that it is caused by interfering with an essential blood function. The massive molecule known as hemoglobin, which has a molar mass of 64,000 g/mol, transports oxygen from the lungs to the blood. Oxygen creates an irreversible bond with the iron atoms found in hemoglobin prior to being transported throughout the body and eventually reaching the cells, where it can participate in the respiration process. Both oxygen and carbon monoxide have extremely similar molecular structures and comparable overall dimensions. Carbon monoxide forms connections with hemoglobin that are approximately 240 times stronger than those formed by oxygen due to its polarity. [For more information on the Lewis acid-base theory and its application to predicting chemical interactions, please see Chemical Connection.] This interaction causes the formation of carboxyhemoglobin, which reduces the quantity of oxygen that the blood can carry. As a result of the accumulation of large amounts of carboxyhemoglobin, the body's cells are unable to absorb the requisite amount of oxygen and eventually begin to die of asphyxiation. This results in the demise of the entire organism. Carbon monoxide is used infrequently in research facilities, whereas the likelihood of carbon monoxide poisoning at home is significantly greater. To ensure that neither user nor anyone else is in danger, it is necessary to execute a comprehensive examination of all potential carbon monoxide sources. Carbon monoxide detectors are an essential safety device that should be installed in all homes with smoke detectors and any equipment that burns gas or fuel. These detectors are reasonably priced.

Other substances may make it harder to breathe over time, but they are not asphyxiants. Inhaling reactive toxicants such as hydrogen cyanide, phosgene, chlorine, ammonia, nitrogen dioxide, or methylfluorosulfate can result in acute pulmonary edema, a potentially fatal condition if not treated promptly. Inhaling reactive toxicants may result in the production of additional toxic effects. These gases contact a person's air sacs when they are inhaled and exhaled. It is unknown how exactly these gases exert their effects, but it is known that they increase the permeability of the air sac-capillary interface. Due to these effects, serum begins to trickle from the capillaries into the air sac after a certain period of time, which can range from a few minutes to several hours depending on the toxicant and the dose. Edema is the medical term for the accumulation of fluid in the air sacs. This condition disrupts the normal exchange of oxygen and carbon dioxide, making it challenging for the affected individual to breathe. If this condition is not treated quickly, the lungs may fill with fluid, causing unconsciousness and death. Avoid situations that expose the user to the risk of reactive chemical toxins. When working with reactive gases, it is essential to always wear chemical shields.

Examples of toxins that can affect development include teratogens and fetotoxicants. There is a wide variety of chemical substances with the potential to inhibit the DNA replication process. The vast majority of these compounds pose risks in the laboratory that are not imminent but rather long-term, potentially leading to serious health issues such as cancer or developmental disorders over time. Carcinogens, which are molecules that cause cancer, are discussed in the section that follows this one. Due to the urgency of the circumstances, it is imperative that the user be familiar with these compounds, even though it is not typical for academic laboratories to possess them. This section provides them in abbreviated form for the users' convenience.

The term "developmental toxicants" refers to any substance capable of interfering with the normal progression of fetal development. Teratogens are substances with the potential to cause malformations and birth defects in a developing fetus. These abnormalities and birth complications can vary in severity. If an embryo is exposed to a teratogen during the first trimester of pregnancy, there is a risk that it will be delivered with a birth defect. Only a small number of substances have been discovered to induce teratogenic effects. The most well-known example of a teratogen is the drug thalidomide, which was discovered to be responsible for limb abnormalities in children born to mothers who took it during pregnancy for purposes such as treating morning sickness and insomnia. Children born to pregnant women who took thalidomide as a sleep aid or to treat nausea and vomiting were born with severe birth defects. There is a minimal chance that a user will encounter this substance while working in the laboratory.

These compounds, known as fetotoxicants, have the ability to affect embryonic development in a variety of other ways, such as causing cognitive impairments, physical deformities, and developmental delays in children exposed to them during pregnancy. Fetal alcohol syndrome [FAS] is a condition in which the infant may be born with a number of possible birth defects. This condition is caused by ethyl alcohol, the most common fetotoxicant. FAS is a disease. The Centers for Disease Control and Prevention [CDC] issued the following statement regarding fatal acute respiratory syndrome [FAS].

Fetal alcohol syndrome [often abbreviated FAS] is one of the most severe complications that can result from drinking alcohol during pregnancy. FASD, also known as fetal alcohol spectrum disorder, is well-established as one of the leading preventable causes of intellectual disability and congenital malformations. A pregnant woman runs the risk of birthing a child with fetal alcohol syndrome [FAS], a condition that can have lifelong effects on a person's physical and mental health and can be transmitted from mother to child. FAS is characterized by atypical facial features, developmental delays, and central nervous system (CNS) dysfunction. People with FAS may have difficulty learning new information, remembering details, focusing for extended periods of time, communicating effectively, seeing clearly, or hearing clearly. Frequently, these issues interfere with an individual's academic and social functioning, leading to challenges in forming relationships, achieving educational goals, and maintaining employment later in life. Persistent symptoms are a hallmark of SARS. It is present in every aspect of an individual's existence, as well as the lives of that individual's family members, affecting their emotional well-being, daily routines, and relationships.

Organ toxins are toxic substances that have an effect on particular organs or physiological systems.

Toxins that primarily affect organs can negatively affect one or more systems and organs of the body. A substantial number of chlorinated hydrocarbon molecules are examples of substances that are toxic to the liver. Carbon tetrachloride and vinyl chloride are both toxic to the liver, but carbon tetrachloride is used more frequently in scientific research. In laboratories, additional chemicals that are detrimental to the liver can be discovered. These include methylene dianiline, dichloroethylene, dimethylformamide, ethanol, and allyl alcohol.

Ingestion of ethylene glycol, a known kidney toxin [nephrotoxic], results in the formation of calcium oxalate crystals, which cause severe kidney injury. This chemical has been linked to a number of high-profile homicides and intentional poisonings, and it is believed that the substance was responsible for these fatalities. Both elemental mercury and mercury salts are exceedingly toxic and accumulate in the kidneys as sulfhydryl-bound Hg^{2}□, which refers to mercury ions (Hg^{2}□) that are chemically bonded to sulfhydryl groups in proteins. Several distinct saline forms of mercury exist. Mercury has been detected in a number of research facilities. Cadmium is another heavy element that, if consumed in sufficient quantities, can be detrimental to a person's kidneys. Chloroform is yet another laboratory-discoverable molecule that is known to cause nephrotoxicity by acting on the proximal tubule of the kidneys.

Neurotoxicants, also known as neurotoxins, are another class of compounds that are considered acutely hazardous. They are substances with the capacity to affect both the central and peripheral nervous systems. We know more about neurotoxic ants than the vast majority of other types of toxic substances because research into their processes has been largely fruitful and has led to the development of numerous pharmaceuticals and other useful compounds. This phenomenon is due to the fact that neurotoxicants are the most prevalent type of toxic substance. This is also because the production of neurotoxic ants has resulted in the development of numerous useful medications and compounds. Pesticides, natural venoms, and nerve poisons are some of the neurotoxicants with the greatest potential for damage. Chemical warfare also uses nerve agents. The law restricts the use of chemical warfare weapons, and even when legal, these compounds require expert handling and storage as well as additional security measures to prevent unauthorized access. These are substances that users are unlikely to encounter in the laboratories that they use early in their undergraduate education, but may encounter later in either their undergraduate or graduate education.

While the neurotoxic properties of some organic solvents, such as n-hexane and carbon disulfide, have been demonstrated experimentally, the neurotoxic properties of other organic solvents remain unproven. A person can be exposed to a drug through ingestion, inhalation, or absorption through the epidermis. The effects on the central nervous system (CNS) can produce intoxication, impairment of psychomotor function, vertigo, headaches, and nausea, among other symptoms. In laboratories, neurotoxic ant poisonings are rarely reported, which is a fortunate circumstance. Toxic substances can also harm the respiratory system, cardiovascular system, immune system, endocrine system, epidermis, and other organs, such as the lungs and skin.

RAMP

- Recognize which types of compounds pose a risk to human health by using the categories presented in this section as a guide for users' analysis.
- Determine the scope of the risk based on the probability of exposure via each of the conceivable routes. "Confined spaces" [also known as small laboratories] should be approached with utmost caution when asphyxiants are present.
- When working with compounds that can be inhaled, it is essential to use ventilation hoods, don protective gear such as gloves, and follow safe laboratory procedures in order to reduce the risk of an accident.
- Ensure that users are prepared for any potential emergencies by being aware of the potentially hazardous effects of the substances users are working with. This ensures that users are prepared for any potential problems.

CHRONIC TOXICITY

Five separate cytotechnologists, each working independently in their facilities, were preparing microscope slides. In laboratories without proper ventilation, workers had no choice but to use xylene as a cleaning and staining solvent for cytological and histological slides during the concluding phase of the procedure. Although the severity of each woman's symptoms ranged from very severe to mild, all of the women reported experiencing recurrent migraines, fatigue, chest discomfort, and other symptoms. Despite multiple trips to the hospital, one woman was ultimately diagnosed with xylene poisoning; however, she was left permanently disabled as a result of her condition. Following their relocation to positions that did not require them to be exposed to xylene, the remaining three users' working conditions improved based on the results of a laboratory investigation, and a recommendation to install laboratory equipment (ventilation). This step was done in accordance with the investigation's findings. As the technician returned to work and put on his hood, he noticed that his health was improving.

Persistent Toxins and Laboratory Testing Results

Chronic toxicants are substances with either confirmed or suspected persistent effects, meaning they can cause long-lasting harm to health or the environment. During the course of the users' laboratory activity, users interact with the aforementioned substances. Even though the event described above is indicative of a negative effect that can result from exposure to a persistent toxicant, incidents such as these in laboratories are extremely rare. Normally, it is necessary to receive dosages over an extended period of time, typically months or years; however, this circumstance is extremely rare in a laboratory setting due to the nature of the adverse effects that can result from chronic toxicant exposure. Typically, doses must be administered over a protracted period of time, typically months or years. Over the course of several decades, the overwhelming majority of laboratories have employed minuscule amounts of numerous chemicals. Due to this, there have only been a very small number of instances in which laboratory exposures have resulted in permanent injury. On the other hand, a few epidemiological studies have demonstrated that chemists as a group have an increased risk of dying from cancer and suicide; the reasons for these observations are not entirely obvious.

In the same manner as in this section, when we discussed acute toxicants, this topic is organized according to the various types of chronic toxicants. On the other hand, experience with these chronic toxicants has been gained through "industrial exposure," in which exposures are frequently much greater than in the laboratory.

This sort of experience has been advantageous. Despite this, it does not follow that the laboratory treatment of persistently hazardous substances is risk-free. When performing any task that involves the management of chemicals, users should take every precaution possible to reduce their exposure. Following is a list of numerous classes of chronic toxicants, as well as examples of chronic toxicants that may be present at a research facility. Carcinogens, also referred to as compounds with the potential to cause cancer, are discussed.

- Recognize which categories of substances pose a risk to human health by basing users' analysis on the categories presented in this section and using them as a guide.
- This enables users to determine which types of substances pose a threat to human health. Users are able to identify the categories of compounds that pose a threat to human health as a result.
- Compute the likelihood that users are exposed in any of the numerous possible ways in order to determine the level of risk users are exposing themselves to. When the encircling environment contains substances that have the potential to induce asphyxiation, "confined spaces" [also known as small laboratories] should be approached with extreme caution.

- It is essential to use ventilation covers when working with inhalable substances. In order to reduce the likelihood of laboratory mishaps, it is also crucial to don protective equipment, such as gloves, and adhere to safe laboratory procedures.
- Users are responsible for educating themselves on the potential adverse effects of the medications they are working with in order to be prepared for any emergency that may arise. This is the most effective method for preparation. This action places the user in a position where they are prepared for any future challenges.

Learn the Fundamental Concepts Necessary for Understanding Chronic Toxicity.

Repeated low-dose exposure to a chemical over an extended period of time results in the development of systemic adverse effects. This definition is the meaning of the phrase "chronic toxicity." Such exposure could potentially cause persistent poisoning, malignancy, or some other adverse effect. The development of chronic toxicity may result from repeated exposure to a chemical at a dose deemed safe. On the basis of acute toxicity alone, chronic toxicity cannot be accurately predicted. Chronic toxicity is much more difficult to identify and study than acute toxicity, which is one of the primary reasons why we know so little about it. Another reason is that the onset of chronic toxicity symptoms may take much longer. Long-term assessments of toxicity are also prohibitively expensive. It is common for the symptoms of chronic poisoning to develop gradually and subtly, and it may be challenging to establish the causal relationship between exposure and illness after a considerable amount of time has passed. Other chronic diseases or the body's gradual decline can obscure the links between symptoms and exposure.

Even though it is possible to infer chronic toxicity from an exposure based on coincidental observations, it is an entirely distinct problem to establish a causal relationship between the exposure and the outcome. Occasionally, extensive and expensive epidemiological studies of large populations reveal prolonged toxicity from chemical exposure. These investigations were conducted on large populations. This work is performed so that statistical analysis can establish associations between exposures and outcomes and generate probabilistic inferences based on these correlations. This is done to allow statistical analysis to establish associations between exposures and outcomes. In general, chronic animal studies can be used to identify the prevalence of chronic toxicants. Subsequent studies, which may be linked to human exposures and outcomes, can shed light on the nature of these toxicants, particularly by examining how they affect human health and the mechanisms through which they operate.

The Role of Epidemiology in Assessing Chemically-Induced Disease

The evolution of the human organism over millennia has resulted in the development of a vast array of protective mechanisms. These defense mechanisms enable the majority of individuals to metabolize the various substances that enter our bodies through our food, drink, epidermis, and lungs in quantities and concentrations that are insignificantly low. Our bodies contain these chemicals in diverse forms. As a result, it is possible for us to be exposed to potentially hazardous compounds at low concentrations, and even occasionally at moderate concentrations, without experiencing adverse effects. This phenomenon is due to the fact that our bodies have the capacity to adapt to a variety of hazardous chemical exposures. This phenomenon is due to the adaptability of the human body. These are handled by the body's defense mechanisms, which either eliminate them or store them in the body's tissues, depending on the appropriate course of action. The following section provides additional information on these metabolic processes.

Think about how a chemical that causes chronic toxicity might affect the user. It is essential to remember that there are millions of chemicals, of which only a tiny fraction have been studied; however, it is already known that some of these molecules exhibit severe chronic toxicity. There is a possibility that certain substances accumulate in the body over time. This shows that the body retains the chemicals and that their concentration rises with continued exposure. In most cases, the body's defense mechanisms can withstand prolonged exposure to low concentrations of toxins for an extended period of time. However, if the toxicant level is increased in the future, the user's body's protective systems may fail or become overburdened due to the additional stress. When the concentration of certain of these compounds reaches a certain threshold, chronic exposure is going to result in injury.

This is because some of these substances are extremely dangerous. The same is the case with substances whose level of toxicity is predetermined. Chronic toxicity follows the same dose-response curve as acute toxicity, but it takes significantly longer to reach this stage at low concentrations. The delay is due to the fact that the toxicant must reach a predetermined toxic level in the body or cause accumulated damage to cells, tissues, or organs. Acute and chronic toxicity follow the same dose-response curve. Acute and chronic toxicity have the same type of dose-response curve. Please consult Chemical Connection for additional information. The body has the capacity to dissolve, store, and then eliminate the substance. [77]

A threshold is attained when the dose of a chronic toxin approaches levels that are capable of producing adverse effects [Figure]. This is when chronic exposure begins to manifest its effects. It is essential to remember that the manifestation of the effects of chronic exposure may take years or decades. This phenomenon is something that cannot be overlooked. After prolonged exposure, the effects begin to accumulate and become more pronounced, and ultimately, non-obvious symptoms manifest. Because we experience a variety of aches, pains, and other types of distress throughout our lives, we might not immediately associate an apparent symptom with a serious issue. This is because we experience aches, pains, and other forms of distress throughout our lives. Eventually, however, the symptoms become glaringly apparent, and in some cases, they may be painful or plainly unpleasant. In most cases, we seek medical attention at this stage, although a physician may have difficulty determining the cause of the illness and must treat it based on the observed symptoms.

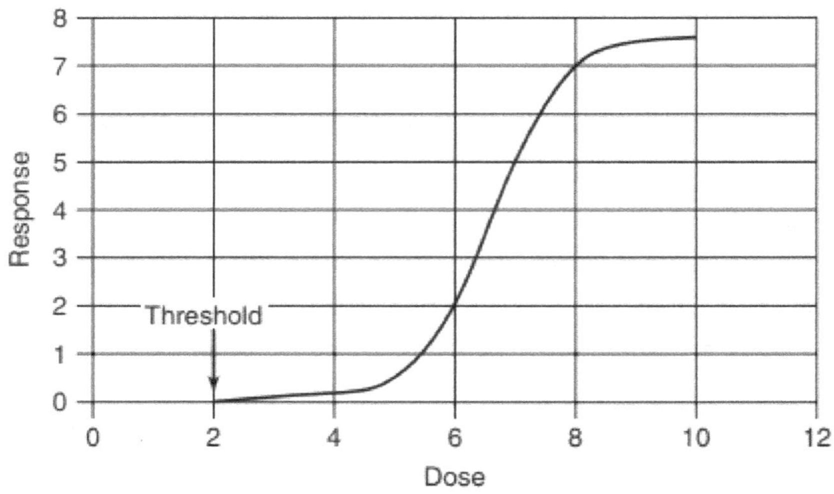

FIGURE: Dose–Response Curve for Chronic Toxicity.

This is an example of a dose-response curve applicable to any substance with a potential for chronic toxicity. On the x-axis of a graph, the amount of medicine that is inhaled or ingested may be indicated. In this example, the threshold is somewhere around the number 3, below which there are no discernible effects. The occurrence of a particular disease or cause of mortality is represented along the y-axis. Even though the presence of a smooth line may provide the impression that there are numerous data points, it is possible that the curve can only be estimated using a limited data set.

Dangers Linked to Prolonged Toxicity

As previously stated, implementing essential safety measures in a laboratory poses a low to moderate risk. However, users are still at risk for prolonged chemical exposures, even if the chemicals in question were brought in from outside the laboratory.

Numerous medical professionals believe that the three factors of heredity, lifestyle choices, and age contribute significantly to the development of chronic diseases. Due to evolution, a variety of biological defensive mechanisms have developed, resulting in our bodies exhibiting strengths in some areas and limitations in others. Consequently, each of us has a genetic makeup that is unique and distinct. This evidence demonstrates that not everyone reacts the same way to prolonged exposure to a hazardous substance. There is a chance that some individuals are able to avoid the potential adverse effects, while others experience symptoms much sooner. As a direct consequence, there are undoubtedly a variety of responses to an exposure.

Users' lifestyle choices, such as whether or not users smoke, drink, or consume excessively; skip exercise; misuse drugs; or engage in other risky behaviors, have a substantial effect on the likelihood that users develop chronic toxicity or a chronic disease over time. In the United States of America, smoking is the most significant risk factor for diseases and conditions that can be avoided by adopting a healthy lifestyle. The Centers for Disease Control and Prevention [CDC] has reported the following information:

The negative effects of smoking affect virtually every organ in the body, increasing a smoker's risk for a wide range of diseases and impairing their overall health. The negative health effects of smoking are responsible for approximately one out of every five fatalities that occur annually in the United States, or an estimated 438,000 deaths per year. Tobacco smoking is the leading preventable cause of mortality in the world, accounting for more deaths annually than the Human Immunodeficiency Virus [HIV], illegal drug use, alcohol consumption, motor vehicle accident injuries, suicides, and homicides combined.

In addition to the natural process of aging, these genetic and lifestyle factors further complicate the detection and investigation of chronic toxicity, which is already challenging, particularly in individuals with a history of smoking or exposure to other toxic substances. Recognizing and comprehending chronic toxicity, although it is frequently difficult, is not an exercise in futility and can be accomplished.

Mercury is a Toxin with Potentially Long-Lasting Effects exposed in the Laboratory.

There have only been a handful of reports of laboratory workers developing chronic toxicity. The majority of these reports are regarded as historical case studies from a time when chemists possessed a limited understanding of the dangers presented by chemicals. Due to the fact that it readily vaporizes and poses a substantial risk when inhaled, mercury has been one of the most significant problems throughout history. At the very least, historical evidence suggests that this has been the case in the past. Alfred Stock was recognized as a leading figure in the development of boron hydride chemistry as a consequence of his forward-thinking research in the early 20th century. In these systems, elemental mercury was used as a key component, and it was most commonly found in manometers, which are pressure-measuring devices. Additionally, elemental mercury was used to manufacture photographic film. Stock's invention of the apparatus and the glassware allowed him to work with air-sensitive compounds.

Stock developed acute mercurialism after being exposed to mercury for an extended period of time while working in poorly ventilated facilities that contained mercury pools and spillage. As soon as he realized that his condition was the result of mercury toxicity, he embarked on a mission that would consume the remainder of his life: he would document the symptoms of this disease and warn other scientists about the dangers of mercury exposure. The symptoms included a decline in intellectual ability, memory loss, speech impediments, and a variety of nerve disorders. Stock was able to make significant discoveries not only in boron hydride chemistry but also in arsenic and phosphorus chemistry, as well as the chemistry of silicon hydrides, despite these obstacles. In addition, Stock was able to combine his understanding of boron hydride chemistry with his understanding of silicon hydride chemistry. W. N. Lipscomb, who was awarded the Nobel Prize in 1976 for his work on boron hydrides, referenced Sidgwick in his Nobel Laureate address. In 1976, Lipscomb was awarded the Nobel Prize for his research on boron hydrides. According to Sidgwick, "all statements made about boron hydrides prior to 1912, when Stock began working on them, are false." Lipscomb was awarded the 1976 Nobel Prize in Chemistry in recognition of his work on boron hydrides.

Even though the majority of modern laboratories have taken significant measures to reduce mercury use, it is still possible to come into contact with mercury while working in a laboratory. Despite the fact that the preponderance of modern laboratories has taken these measures, this remains the case. Mercury metal is a dense liquid used in a variety of scientific instruments, including thermometers and manometers. Additionally, mercury is used in electrical switches, thermometers, and fluorescent light sources. In the event that broken thermometers, other mercury-containing equipment, or mercury spills are not cleaned up as soon as they are discovered, there is a chance that individuals are chronically exposed to mercury over an extended period of time. Due to the viscosity of liquid mercury, it is able to seep into fissures and cracks, making the extraction procedure extremely difficult. If a person comes into contact with or observes a mercury release, there are some who believe the spill is not severe enough to warrant cleanup. Mercury could still be present even if it "disappears" when it enters these low-lying areas and becomes dispersed throughout the flooring and tiling. Due to the fact that mercury has a low but significant vapor pressure, it is easy to disseminate into the air within a laboratory, although inhaling mercury is fatal. As a result, chemists run the risk of being exposed to mercury through inhalation for extended periods of time. This is the ideal recipe for a disease that induces gradual poisoning.

According to a 1972 report on mercury contamination in chemical laboratories, a number of university students, professors, technicians, and workers were discovered to have chronic mercury toxicity or elevated urine mercury levels. 12 It was determined that a student who complained of persistent fatigue, migraines, and mental sluggishness may have been exposed to mercury in his laboratory. This made it clear that this was what caused the student's symptoms. The student's prior experience included working with mercury. During the course of the investigation, it was determined that not only did his urine contain excessive levels of mercury, but so did the air in the laboratory. Even though their urine contained abnormally high levels of mercury, only three of the twenty-eight laboratory workers reported experiencing symptoms. Mercury that had leaked from damaged thermometers and manometers over their many years of service was responsible for the widespread contamination of a number of laboratory rooms. This was the fundamental reason for the pervasive pollution. As a direct consequence of the decontamination procedure, the mercury concentration in the laboratory's air was reduced to safe levels. The urine concentrations of individuals who had been exposed to the toxin quickly returned to normal levels.

Additionally, in situations where it is practicable to do so, decreasing the amount of mercury-containing equipment used can reduce the risk of mercury poisoning. When feasible, replace mercury-containing equipment with alternatives that do not contain mercury. Thermometers that do not contain mercury are examples of such alternatives.

Utilization of Chronically Toxic Substances in the Workplace

When working for an extended period of time in a laboratory, the likelihood of being exposed to a large number of chronic toxicants is low. This was covered in the preceding section. Nonetheless, even if this proves to be the case, the implementation of modern safety standards substantially mitigates the severity of this risk. The user must know the OSHA rules and the toxicants that fall under this category, as the user may one day manage labs or factories. The OSHA Laboratory Standard [29 CFR 1910.1450] requires laboratory workers and their supervisors to be aware of the potential dangers and exposure hazards posed by chronic toxicants. This law pertains to all compounds that cause long-term harm. The determination of maximal safe exposure levels falls under the purview of various public and private institutions. OSHA is responsible for enforcing restrictions known as permissible exposure limits [PELs], for instance. The American Conference of Governmental Industrial Hygienists [ACGIH] endorses TLVs, also known as threshold limit values, with a resounding endorsement. IDLH stands for "immediately dangerous to life and health," and the National Institute for Occupational Safety and Health [NIOSH] has a compendium of IDLH levels. In addition to those, there are also other relevant standards. The following section, 6.2.2, provides an in-depth discussion of the numerous regulations, suggestions, and acronyms pertinent to both acute and chronic toxicants.

RAMP

- Determine the risk level based on the likelihood of exposure via multiple routes. This is the first stage of the RAMP procedure.
- Identify the chemical components that constitute chronic toxins.
- Because an episode of chronic toxicity lasts longer than an episode of acute toxicity, it is less important to prepare for unanticipated events. The greatest "preparation" is knowledge of the signs and symptoms of chronic exposures, which necessitates the actions outlined in the preceding section: recognize, assess, and reduce the risk.

BIOLOGICAL HAZARDS AND BIOSAFETY

One of the researchers was about to conduct an analysis on human serum samples. She used lengthy Pasteur pipettes to transfer serum samples from one receptacle to another. After completing the investigation, the scientist disposed of the used pipettes by placing them in a metal pan and then sterilizing them in an autoclave. Another researcher disposed of his unused pipettes by placing them in the metal pan; however, when he lifted the lid of the autoclave pan to place his pipettes in it, he was pricked by one of the pipettes in the box. He found out that the pipette had previously transmitted human serum. After washing the wound with detergent, he went to the hospital to have his injury examined. The physician informed him that the only thing he could do was endeavor to determine whether the serum contained infectious components; however, this was not a simple task. If the situation is not the case, he needs to be closely monitored to determine if he develops symptoms of an infectious disease or any markers indicating he has developed immunity to viruses. If this is not the case, he must be monitored attentively.

Pathogens, which are agents that can cause disease, include both bacteria and viruses.

There is always the possibility of exposure to biological agents, also known as microbiological agents, when operating in a laboratory. These agents have the potential to induce disease. Workers in clinical laboratories, hospitals, and physicians' offices risk coming into contact with microbiologically infectious substances. The preceding comment indicates that exposure to these microbiologically infectious elements is possible. These possibilities can occur in a variety of contexts. The Occupational Safety and Health Administration [OSHA] estimates that approximately one-fourth of these positions are available in academic laboratories. Whether used for instruction or research, both chemistry and microbiology laboratories run the risk of containing biological agents, which can include bacteria, viruses, and fungi that may pose health risks to laboratory personnel and the surrounding environment. In addition, microbiological organisms are virtually ubiquitous; however, the good news is that only a tiny fraction of these organisms pose a threat to human health and are pathogenic. Although this section provides an introduction to the potential hazards posed by microbiological agents in the laboratory, it does not cover all pertinent information.

Despite the existence of other classes of pathogenic microbiological agents, this is the case.

- Viruses are subcellular and submicroscopic organisms with a nucleic acid [DNA or RNA] gene core that can assume a variety of configurations. This core is surrounded by a protective protein and is capable of being encapsulated in a lipoprotein membrane. It is a widely acknowledged fact that viruses are not considered living organisms.
- Their width can be as narrow as 0.02 micrometers, while their length can reach 0.3 micrometers. To ensure their continued existence, viruses must infect a host cell and then utilize the cell's energy supply and metabolic processes in order to proliferate when environmental conditions are favorable. Independent existence is unthinkable for viruses. Therefore, antibiotics should not be used to treat infectious diseases caused by viruses. Vaccinations are a method of disease prevention that can be used to protect against conditions caused by particular viruses.
- The vast majority of microorganisms have a diameter between 0.5 and 1 micron and a length between 2 and 5 microns. Their lengths vary between 2 and 5 micrometers.
- Because their cell walls contain components that react with different dyes, they can be classified as gram-positive [retaining the dye] or gram-negative [not retaining the dye] based on whether or not the dye is retained. This is due to the fact that their cell walls contain specific components that react with various pigments. Bacteria are able to replicate within a host under favorable conditions and in the presence of all the essential components of a conventional cell. It is feasible to cultivate "colonies" of bacteria on suitable solid culture media. Antibiotics are frequently prescribed by physicians to treat bacterial ailments.

Risks of Bloodborne Pathogens Associated with Human Blood Specimen Handling

Because of the nature of infectious diseases, the dangers presented by pathogens are distinct from those posed by chemicals. Since infectious agents have a tendency to proliferate once they have infected a host [in this instance, the user], even a seemingly mild exposure could result in a severe illness if left unchecked. This phenomenon is because infectious agents proliferate after infecting a host. Users must develop a strategy to protect themselves from exposure to these agents, just as users must develop a strategy to protect themselves from potentially hazardous substances, which may include using personal protective equipment, practicing good hygiene, and following safety protocols established by health organizations.

The Occupational Safety and Health Administration [OSHA] issued a regulation in 1991 titled "Occupational Exposure to Bloodborne Pathogens." Since then, the Occupational Safety and Health Administration [OSHA] has reviewed and revised this regulation twice, first in 1999 and then again in 2001. Even if there are additional infectious substances, we focus on bloodborne pathogens, also known as BBPs. These are the infectious agents most likely to cause disease transmission. This is due to the fact that bloodborne pathogens are the primary focus of this event.

There are multiple potential routes of exposure to infectious agents in the laboratory, just as there are multiple potential routes of exposure to chemical hazards. Additionally, there are numerous possible routes of exposure to chemical hazards. Exposure routes include inhalation, injection, cutaneous contact, and ocular contact.

Because secure laboratory practices prohibit eating, drinking, smoking, applying cosmetics, taking medications, and other similar activities while in the laboratory, ingestion should not play a significant role as a route of exposure in the laboratory. Pipetting with one's mouth, which was prevalent in laboratories in the early 20th century, is no longer permitted. It was more prevalent at the turn of the century.

- If a user has wounds, scrapes, abrasions, dermatitis, or other skin disorders, dermal exposure can be particularly harmful. These skin conditions are susceptible to splashes, aerosols, and other accidental exposures, and they create entry sites for microbiological agents. If a user has any of the following skin conditions, dermal exposure can be especially hazardous:
- Accidental droplets or discharges of infectious materials that enter the eyes of those who are exposed to them may cause potentially fatal infections. Research has demonstrated this to be the case.
- Inhalation is the primary route of exposure to infectious pathogens that have become airborne due to a variety of laboratory processes. In the following sections of this essay, several containment strategies are examined.
- Accidental injection during the handling of potentially infectious materials, such as blood samples, is a serious concern; consequently, measures to reduce accidental exposures from "sharps" are proposed as a means of mitigating this risk. This is due to the fact that accidental exposure to sharps is the leading cause of inadvertent injections.

Taking Precautions to Guard against Bloodborne Pathogens and Other Infectious Materials

Infections with the hepatitis B virus [HBV] are capable of causing extensive liver injury. The greatest defense against HBV infections is vaccination, which is extremely encouraging news. The laboratory staff who regularly handle blood samples strongly encourage every member to receive the HBV vaccine. This precaution protects them from the virus's effects, even if they unknowingly encounter it

To protect oneself from exposure to BBPs in general, however, it is necessary to adhere to specific laboratory practices. The user may encounter the term "universal precautions," which implies that all human specimens [body fluids and tissues] should be treated as if they were infectious. Nevertheless, the user should perform these actions.

The universal precautions include a range of safety measures, each of which is detailed in the following paragraphs. [78]

RAMP

- If a user handles human specimens or other tissues or secretions from living organisms, the user runs a very real risk of contracting infectious agents, which can cause potentially fatal diseases.
- Determine the extent of the potential hazard posed by the laboratory procedures.
- Take the precautions outlined in the Bloodborne Pathogen Standard, or BMBL, to reduce users' risk of exposure to a bloodborne pathogen. If the user wishes to prevent infections caused by specific viruses, the user should investigate the possibility of receiving vaccinations.
- It is of the utmost significance to be familiar with the best course of action in the event of an unanticipated exposure.

RECOGNIZING LABORATORY HAZARDS: PHYSICAL HAZARDS

SUMMARY

In this chapter, we have studied the numerous approaches developed by the chemical community and other organizations to communicate the numerous hazards associated with chemical exposure. The beneficial news is that there is a wealth of information on the dangers of many of the compounds users are likely to encounter in chemistry labs. The unfortunate news is that it often takes a bit more than "common sense" to understand the hazards involved. The beneficial news is that there is a wealth of information on the hazards of many of the chemicals users are likely to encounter in chemistry labs. In some instances, it may be difficult to locate the desired information about chemical hazards; therefore, be prepared to exert effort in researching safety data sheets, academic articles, and reliable online resources. After completing this chapter, students have a deeper understanding of how to locate and comprehend a variety of safety-related information because they have gained this knowledge. The Internet has made this "finding" marginally simpler, but "interpreting" it correctly still requires comprehension on the users' part! This section is the first of three chapters discussing the central concept of safety, which is being aware of potential threats. This section discusses the potential pathways that could contribute to exposure and the methods for blocking these pathways to prevent exposure. The principles of comprehending labels, signs, symbols, words, and other information sources are taught to laboratory personnel to train them in the essential skill of hazard recognition, which is required of all laboratory personnel. This is because laboratory users must be able to identify potential dangers. In this section, users will discover a collection of resources containing information regarding the hazards posed by chemicals. It provides a description of these resources as well as potential applications for their use. This section provides an overview of readily accessible information sources and discusses the dangers posed by chemicals. With this information, users are able to identify the dangers posed by chemicals and take the necessary precautions to safeguard themselves.

It is possible to increase a person's awareness of the need to avoid chemical exposure by increasing their knowledge of the toxicity of the chemicals themselves, which can also help them comprehend other sources of exposure besides ingestion. Consuming laboratory chemicals is not something that the majority of chemistry students would ever contemplate doing; however, doing so can help students become more aware of the risks involved. Even though the majority of chemical exposures, if they occur at all, are likely to be fleeting in academic laboratories, certain career paths pose the risk of prolonged exposure to small quantities of chemicals. Other industrial environments also pose this risk. Consequently, it is in users' best interest to have a thorough understanding of the numerous ways in which chemical exposure can be detrimental to users' health. This concept should be expanded to encompass both carcinogens and biological agents. Laboratory safety requires the capacity to recognize potentially dangerous situations.

To ensure one's safety, it is necessary to develop the ability to recognize potentially hazardous aspects of one's physical surroundings. In their laboratories, chemists utilize a vast array of compounds with a broad range of distinct physical properties, as well as an assortment of odd pieces of equipment. Chemists of all levels of experience, including students, have the potential to cause countless accidents in the laboratory if they lack sufficient knowledge of the chemicals and apparatus they are employing. This chapter covers lab injuries and how to avoid them.

CHAPTER

04

TECHNICAL EVALUATION OF A CHEMISTRY LABORATORY

Safety Practices and Observations in the Laboratory

Everyone agrees that ensuring all safety standards are adhered to at all times in a laboratory would result in an exceptionally low risk level. There must be strict standards for monitoring and inspecting all laboratories, irrespective of whether they are located in educational institutions or manufacturing facilities. [65]

In light of these considerations, we have compiled a condensed summary of the observations and preventative measures that, in our opinion, are necessary to guarantee the laboratory's continued safety. [66]

The provision of fume hood containers is to ensure the safest possible administration of dangerous or hazardous substances. [67]

During a routine laboratory inspection, we found that these fume chambers were storing various chemicals. As a direct result, the interior space cannot be effectively utilized for conducting reactions or processing additional substances, which compromises safety and efficiency in the laboratory environment. [68]

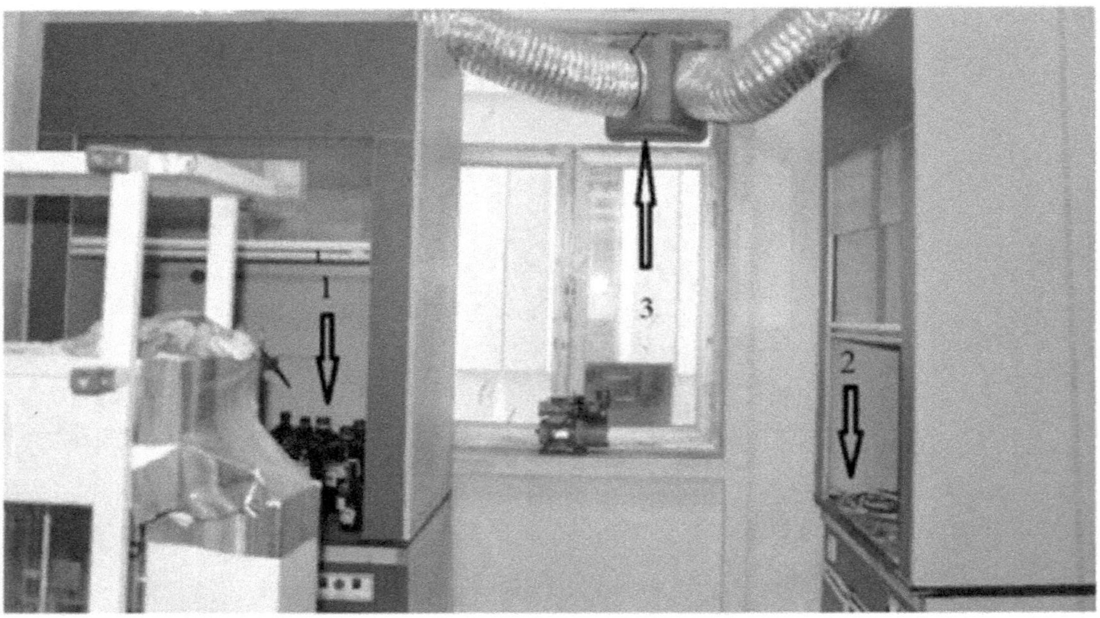

FIGURE 1: Fume hood cabinets in the laboratory

Figure 1 depicts the presence of specific electrical wires inside the fume hood and the connection of ventilation tubes using a single common receptacle, which are two additional noteworthy observations. Because chemical vapors can travel from one fume hood to another if one is turned off, connecting ventilation ducts with a single common outlet entails the potential for extremely high levels of risk. In addition, ventilation outlets must be located at a safe distance from any routes and cannot be constructed in the manner depicted in Figure 1. [69]

FIGURE 2: Unsafe gas cylinder position in the laboratory

The addition of numerous gas vents to workbenches makes conducting low-risk experiments on these benches much simpler. On the other hand, the gas was administered through the use of gas cylinders [as depicted in Figure 2], which entailed a high degree of danger. It is strongly recommended that these gas cylinders be stored outside the laboratory in a sheltered and shady area and that the gas flow inside the laboratory be controlled by installing regulators with indicators. These two precautions must be followed. [70]

To reduce the risk of potential fires, we believe that the use of exposed flames in this laboratory should be severely limited, if not prohibited entirely. This would help reduce the overall risk of fire. In such a scenario, heating mantles and using heated dishes are viable options. [71]

In addition, a used water container was discovered on the seat [see Figure 2], which is a violation of the safety measures because eating and drinking are never permitted on an airplane. [72]

Safety Instructions

Detailed instructions were provided for the numerous laboratory procedures. However, as shown in Figure 3a, the majority of laboratory workers believed that the instructions were excessively detailed and ineffective, leading to confusion and decreased compliance with safety protocols. To maximize the benefits of following instructions, as depicted in Figure 3b, we utilized a variety of safety instructions in the laboratory. Our objective was to maximize the advantages of following the instructions. We provide easily comprehensible and informative visuals and indicators to accompany these instructions. [73]

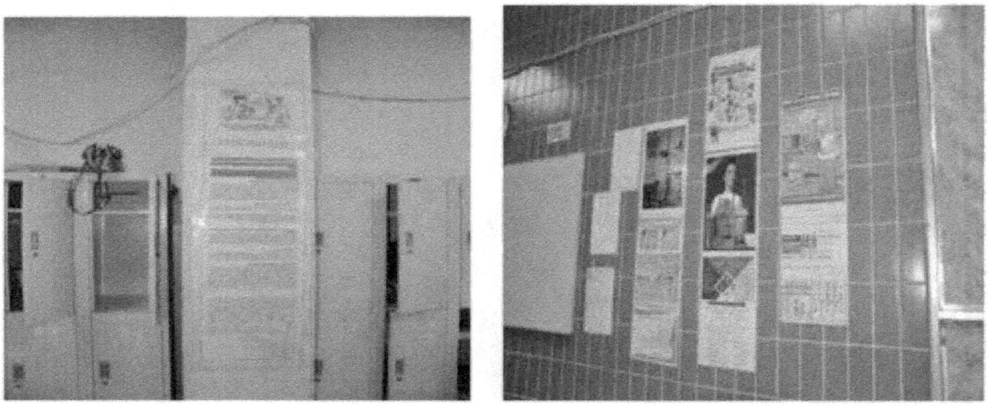

[a] and [b]

[FIGURE 3: a] Previous safety instructions with no signs and figures b] Current safety instructions supported with figures, signs

In the event of an emergency, both a lavatory and an eyewash station are available. As illustrated in Figure 4, the senior laboratory demonstrator ensured that the students had access to a man-made eye cleanser and a shower in the event of accidents involving their eyes or bodies. [74]

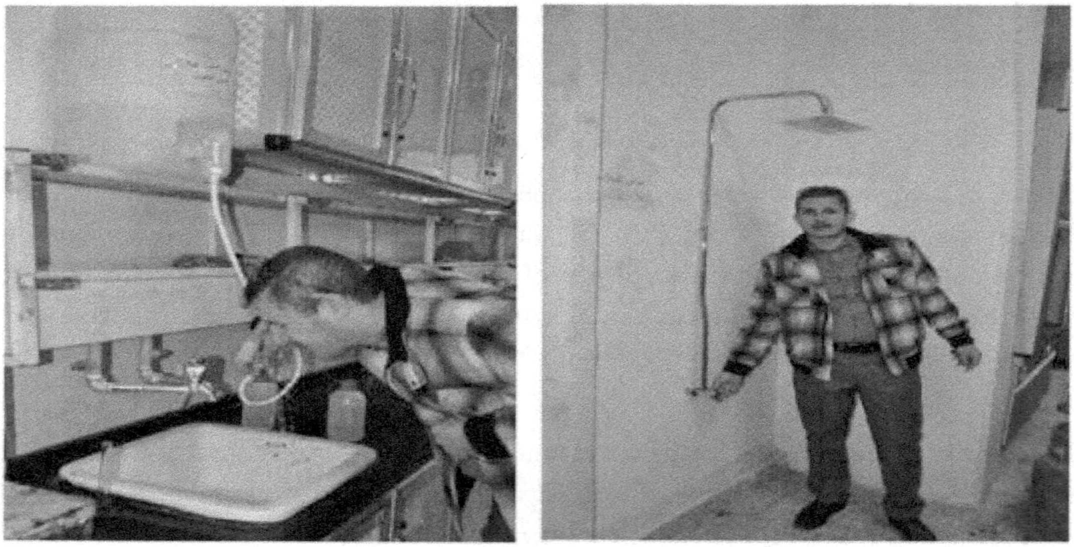

FIGURE 4: The current emergency eye washer and shower

Currently, this piece of equipment can satisfactorily meet the requirements. Figure 5 illustrates the value of deploying technology that is more user-friendly and compliant with international standards. Figure 6 depicts a first aid kit that can be used to treat various minor injuries, including wounds and burns. To satisfy the needs of those who have been injured while performing their jobs in the laboratory, it is necessary to establish guidelines. [75]

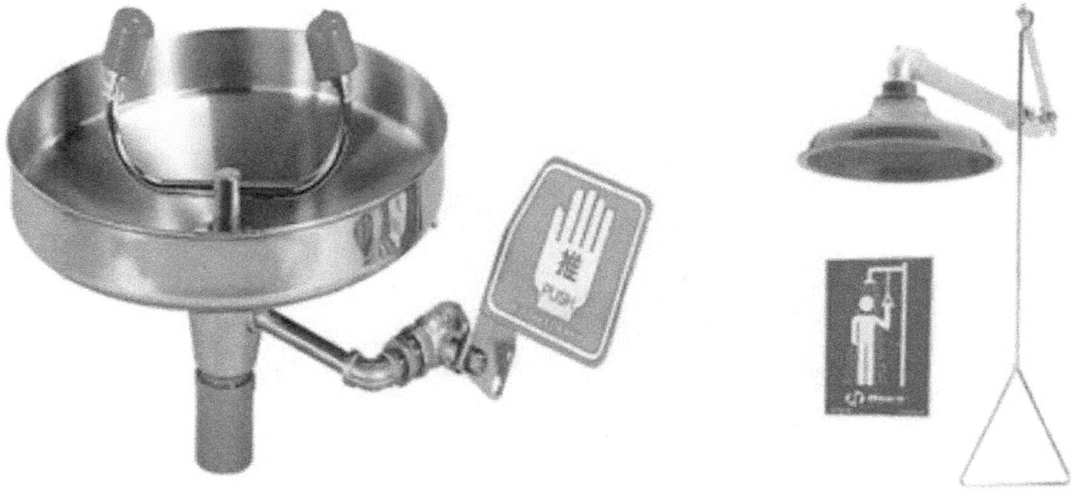

FIGURE 5: Left, a standard eye washer. Right, a shower with an easy-to-use handle

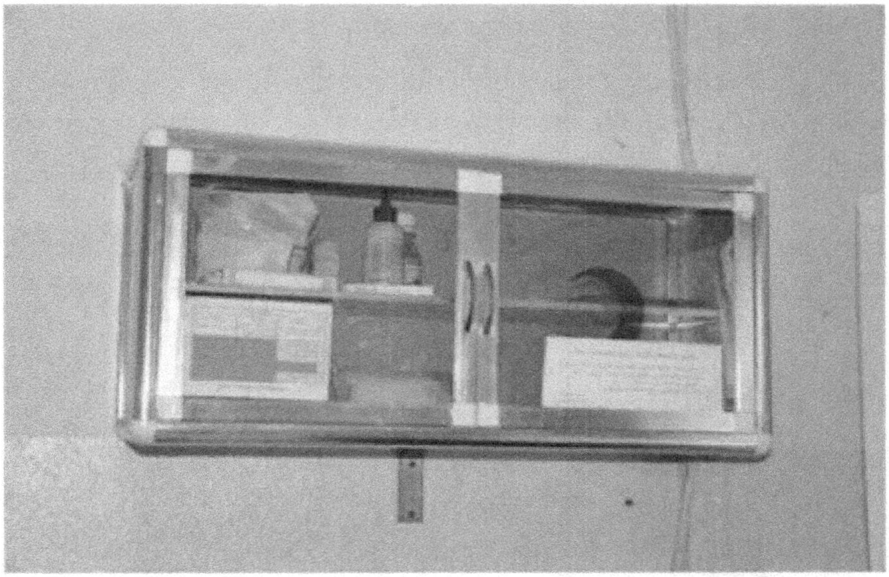

FIGURE 6: The first aid kit mounted on the wall

There was firefighting equipment in the laboratory, such as fire extinguishers [Figure 7]. In contrast, users should not place them on a perch due to the hazards involved. We believe hanging them on the walls is the most effective method to display them; at least two should be displayed inside the laboratory, and one should be displayed outside. In addition, it was discovered that the laboratory in question lacked a designated emergency exit [as depicted in Figure 8], a violation of the safety regulations. As a consequence, we recommend that the laboratory be equipped with a rescue exit that features an outward-swinging door. [76]

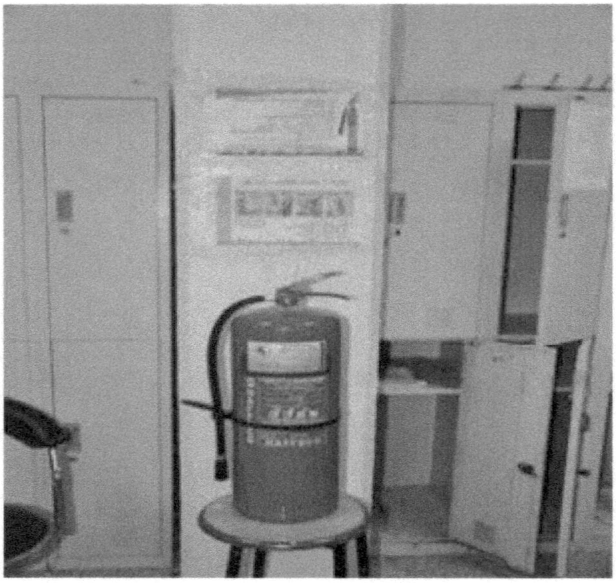

FIGURE 7: A fire extinguisher improperly placed on a stool.

In addition, it was discovered that the laboratory in question lacked a designated emergency exit [as depicted in Figure 8], a violation of the safety regulations. As a consequence, we recommend that the laboratory be equipped with a rescue exit that features an outward-swinging door. [77]

FIGURE 8: The absence of an emergency exit in the laboratory

SUMMARY

Chemistry laboratories, due to their persistent exposure to potentially hazardous chemicals, are classified as high-risk workplaces. It is exceedingly unprofessional and negligent to handle any of these compounds, some of which have a high-risk score, without taking into account the possibility that they may cause injury to people or damage to property. On the basis of a comprehensive evaluation of one of the chemistry laboratories at our educational institution [the Department of Chemistry in the College of Education for Women at Baghdad University in Iraq], we have deduced a feasible safety practice to provide a reasonable level of protection for both people and facilities. Our objective is to achieve it in a manner that safeguards people and property to the utmost extent possible. This evaluation was conducted to guarantee the protection of both individuals and property. In chemistry laboratories, safety procedures must be observed at all times, and a safety officer must conduct routine inspections to ensure compliance with safety regulations. Following is a summary of our evaluation, our observations, and our recommendations for strengthening the current safety measures at this chemistry laboratory.

CHAPTER

05

SAFETY EQUIPMENT IN CHEMISTRY LAB

This list may not include every conceivable guideline for ensuring the safety of a laboratory, but it does provide a comprehensive overview of the concepts. Monthly laboratory quality inspections are conducted throughout the academic year. Every laboratory must conspicuously display a copy of the "Safety in Academic Chemistry Laboratories" pamphlet published by the American Chemical Society. Please consult the enclosed pamphlet for additional instructions and suggestions. It is possible that laboratory supervisors may request that the user implement additional safety measures. [79]

Equipment for Emergencies

- Fire extinguishers should be placed in strategic locations near exits, and they should be of the appropriate type and size.
- Laboratory spill kits should be easily accessible.
- Users of the lab conduct routine inspections, and Environmental Health and Safety conducts annual inspections of the safety showers and eyewash fixtures. [80]

Hoods for Ventilation and the Collection of Fumes

- The hood is functioning within what are considered "normal" standards, and the alarm system is currently engaged.
- There is no obstruction to the air movement within the hood, which denotes that the hood is not congested, that objects are stored behind the 6-inch line, that bulky items are placed on blocks, and that the baffle area is clear.
- A hood is not to be used for storage unless it is properly designated and tagged 'for storage solely'.
- A hood or an adequate gas trap is to be used when hazardous gases or fumes are created.
- A particular hood is needed when working with perchloric acid.
- Filters are installed on vacuum pumps, or vents are connected to the exhaust system.
- Windows are shut. [81]

Security Personnel

- Even if users are not working with liquids, they should wear side-shielded safety eyewear.
- It protects users' eyesight. Eye protection that is resistant to both spills and impacts must be worn when working with liquids.
- Visitors must wear safety glasses.

- Proper gloves and clothing [long pants, closed-toe shoes, and a lab coat] must always be worn.
- Gloves and lab coats must be removed before exiting the laboratory.
- Long hair must be restrained. [82]

Chemicals and Compressed Gas Cylinders

- Material safety data sheets [MSDS] are available for all of the chemicals that are utilized in the laboratory.
- Chemicals are separated and stored according to their class [for instance, store acids away from bases, reducing agents away from oxidizing agents]
- Chemicals are kept in a location that is shielded from direct heat and sunlight.
- Flammable liquids should be stored in safety containers that are specifically designed for the purpose, or in a solvent cabinet.
- Chemicals should be stored in appropriate, lidded containers.
- Cabinets containing extremely hazardous chemicals are secured with locks.
- Compressed gas cylinders are secured individually.
- Gas cylinders are capped when not in use.
- Lecture bottles are stored in racks, clamped, or on their sides [83].

Chemical Waste

- Waste must be separated into the appropriate categories, tagged [with the type of material, the quantity, and the date it was added], and deposited in sealable containers.
- Waste cans plainly marked "hazardous waste" or "non-hazardous waste."
- Forms for waste accumulation that are strategically located near garbage cans.
- Disposal requests must be submitted in a timely manner via the Chemical Tracking System [84].

Labels and Signs

- All chemical container labels must be written in English and include the current date.
- Containers for long-lasting reactions [those that require at least one night to complete] must include the structure or formula of the reaction, as well as the initials of the reactants.
- Signs indicating the presence of specific hazards, such as radioactive materials, lasers, or ultraviolet light bulbs.
- All chemical storage cabinets must bear distinct labels [such as "flammable," "acid," and "base"]. [85]

Following signs posted in all labs

- This is the inventory of items that need to be eliminated.
- Eyewash fountain and emergency shower testing and utilization page.
- Procedures for the disposal of hazardous materials.

- Emergency procedures.
- General policy on working in research facilities.
- Guidelines for the safe management of blood-borne pathogens.
- Procedures for the disposal of hazardous materials.
- Emergency procedures page. [86]

Obstructions and Hazards

- Emergency exit unobstructed and unlocked
- Taped area under safety shower and eyewash fountain kept clear
- Aisles and paths unobstructed
- No cords or tubing obstructing or dangling in paths
- Belt guards on pumps
- Apparatus properly set up
- Floor and benchtops clear of spills
- No bicycles in laboratories [88]

Electrical Hazards

- Each item of equipment may have one extension cord.
- Extension cords may not pass through doors or ceiling grids;
- Extension cords must be in excellent condition.
- Fuses or circuit breakers are required to protect extension cables used permanently. [89]

Housekeeping

- It is prohibited to position items such as broken glass, bare needles, chemical spillage, trash, and similar items on the benches.
- Ensure that glassware and other potentially dangerous items are kept away from the bench's edges.
- Ensure that the washbasin drains are covered with mats or screens and are spotless.
- Needles and other pointy objects that require disposal must be stored in a separate, labeled container.
- A defrosted freezer and a spotless refrigerator are required. [90]

Food and Beverages

- Users are not permitted to bring food, alcoholic beverages, coffee appliances, or other similar items into the laboratory.
- Additionally, laboratory water should not be ingested; instead, it should be obtained from water fountains or restrooms. [91]

Safety Equipment

Acquire a thorough understanding of the following pieces of emergency apparatus, including their locations, functions, and proper usage.

1. A wall-mounted storage container functions as the location for the fire extinguisher, which is attached to the wall.
2. At eye level, a bright red fire alarm has been affixed to the wall close to the fire extinguisher.
3. Adjacent to the fire extinguisher is a numbered red case that has been mounted to the wall. Inside, you'll discover a fire blanket.
4. The blanket is used to extinguish any clothing flames that may have already started.
5. In this situation, a person who has just been jolted by electricity may benefit from having the blanket draped over them.
6. The emergency lavatory in the laboratory is conveniently located adjacent to the building's main entrance.
7. If the handle is pulled down, a supply of water is going to descend, allowing the individual to rinse off their entire body as well as their clothing if a hazardous chemical is sprayed on them in large quantities.
8. There are several distinct basins in the laboratory, each with its own drench hose attached.
9. This technique is useful for cleaning up small areas of the body affected by localized chemical spillage. In addition to that, it can also be used as eyewash.
10. The building's eyewash station is located directly beneath the safety shower.
11. Additionally, at least two additional eyewash stations are dispersed throughout the laboratory, each in a distinct location.
12. Each is connected to a portion of the conduit that can be detached separately.
13. When fully operational, it features a number of miniature fountains that eject water towards the heavens. Fume vents are installed along the room's perimeter and at various locations within it.
14. These are used in areas where poisonous or unpleasant vapors are present in order to eradicate those vapors.
15. A phone designed for emergency use. This is a wall-mounted item that can be positioned near the emergency shower. This device also allows users to contact campus police by dialing 911, so users can use it to make calls within the school as well.
16. A first-aid kit is located on the laboratory bench near the building's exterior entrances.
17. The baking soda, or sodium bicarbonate, can be found on the bench near the laboratory's exit. Sodium bicarbonate is also the chemical name for baking soda. Using this product, users can neutralize any acid that may have been released.
18. The spill equipment is kept in a cabinet that has been meticulously labeled and is located within the laboratory. It is designed to be used as a cleaning solution in the event of a chemical spill on countertops or surfaces.

19. Everyone in the building should be conversant with the building's emergency exits and response procedures. In the event of an emergency that requires the user to evacuate the premises as soon as possible, the user should familiarize themselves with all of the exits in the lab and the building.
20. Infrastructure for Urgent Message Transmission Intercom: There is a button and a picture of a phone displayed on the intercom in each of the corridors directly opposite the laboratory entrances.
21. These corridors are on the opposite side of the structure. This line weaves its way directly to the Cerritos College campus safety office. [92]

CHAPTER

06

CHEMISTRY LABORATORIES SAFETY RULES

This chapter provides an overview of the safety ethic and the four safety principles covered in the majority of the book's other chapters. In addition, there are three sections devoted to green chemistry, which suggests that there is a significant link between this new way of contemplating environmentally friendly chemistry and chemical safety. Last but not least, the topics covered in Level 3 focus on a variety of safety issues that, while they may seem less important to users as students, are of the utmost importance to users as scientists once users graduate from college and enter the professional world of chemistry, including proper laboratory practices, risk assessment, and emergency response protocols. [93]

THE FOUR PRINCIPLES OF SAFETY

The acronym for "recognize, analyze, minimize, and prepare" was previously mentioned. Let's examine each of these stages from the perspective of the potential dangers and hazards that could arise at any time.

Users are responsible for understanding and recognizing the hazards associated with the drugs they are working with. This requirement is due to the fact that the primary concept is to identify the hazards posed by the substances, equipment, and methods being utilized. It doesn't seem particularly difficult, does it? The level of difficulty associated with the assignment will unquestionably be directly proportional to the quantity of knowledge the user already possesses. It's impossible to identify the dangers of any one substance because there are millions of them. However, there are two distinct factors that contribute to our success.

To begin with, the overwhelming majority of chemicals can be classified into one or two of a small number of known hazard categories. These categories were determined through extensive research. Extensive research has been conducted regarding these categories. Before a user can completely comprehend these threats, the user must become familiar with the vocabulary and information that characterize the various chemical qualities. This is required so that the user can comprehend them. Consequently, the user will have a comprehensive comprehension of the scenario. What does it mean for something to possess the property of being "flammable"? How truthful is it to describe someone or something as "corrosive" or "toxic"? How do users intend to ascertain if the chemical in question possesses any of these characteristics?

Second, and more specifically, "getting to know users' chemistry" requires that the user read and comprehend the publicly available information about the potential hazards posed by the substance. This information is available online and accessible. This information is accessible via the Internet in various formats. As part of this procedure, users are required to peruse the labels on the container, the Material Safety Data Sheets, reference materials, and internet-based hazard information. In addition, users need to seek the advice of individuals with relevant experience. The second of these suggestions is to conduct an analysis of the hazards associated with the different processes and exposures. This is, without a doubt, the most essential point to remember. As a result, the user must conduct a risk assessment of the dangers associated with the operation of the apparatus and the multiple ways in which the user could be exposed to a wide variety of chemicals during a particular process or reaction. Is it possible for this process to generate enough heat or other forms of energy to start a conflagration or cause an explosion? Is there the potential for combustible materials to be present, which could result in a fire? How likely am I to come into contact with something that could be harmful?

It is crucial not to become complacent regarding the potential dangers that may exist in any location, but particularly in "familiar" environments. If users read this book and make an effort to participate in the chemistry courses they are enrolled in, their ability to identify potential risks can improve significantly.

To effectively implement the third principle, which is to minimize hazards, a person must pay close attention to both the planning and execution of an experiment. Only then will they be able to implement the principle effectively. As a result, it is of the utmost importance that users take all of the necessary precautions for laboratory safety to reduce, eliminate, or obtain control over any hazards associated with their working environment. Conducting a thorough investigation of the potential hazards is the first step to carrying out this task. Personal protective equipment [PPE] such as splash goggles and other safety equipment [such as chemical shelters], along with adequate sanitation practices, are essential components for reducing potential hazards to acceptable levels. Other safety equipment includes hoods designed to protect against the inhalation of hazardous particulates. PPE includes eyewear designed to protect against liquid splashes. Hoods are another item of protective clothing worn to shield the wearer from potential chemical hazards. The purpose of goggles, which are part of PPE, is to protect the wearer's eyes from any potential accidents. Moreover, research must be conducted with awareness of the inherent risks involved. The presence of disorder and disarray in the workplace is the primary contributor to numerous incidents.

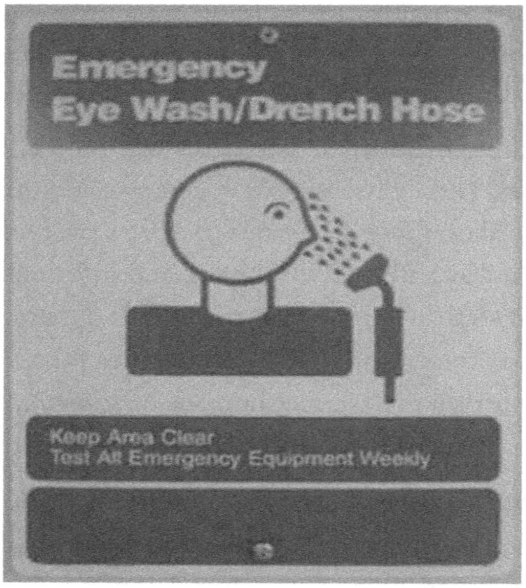

FIGURE: Eyewash Safety Sign. There are usually several signs indicating the location of safety equipment in laboratories. These often have pictograms that indicate the use of the safety equipment.

Even if lab errors and exposures are prevented, it's wise to be ready for them. Therefore, the fourth rule is to always be ready for unexpected events. In a research facility, various unanticipated occurrences may arise. Each and every one of these potential dangers, including but not limited to fires, explosions, accidental chemical exposure, and personal injuries, has been considered at some point. The figure indicates that, to be prepared for unanticipated events, it is crucial to have a functional understanding of the numerous types of readily available safety apparatus and how to operate them. This information is essential for being prepared for unforeseen circumstances. Additionally, users are expected to know when and how a laboratory should be evacuated, as well as the location of the first aid supplies. In a chemical lab, users must learn crisis management first.

The Student Safety Ethic

This is the aspect of safety ethics that the students focus on. In the section that follows, we discuss the safety ethic, which expands the scope of the student safety ethic to include a broad range of issues pertinent to practicing chemists and researchers. This goal is accomplished by comparing and contrasting the safety ethic of the students with that of the school.

The term "ethics" refers to the guiding principles that establish what constitutes ethical behavior or conduct. Having a set of health and safety guidelines to refer to when conducting laboratory work is advantageous because it helps ensure that students understand the importance of safety protocols and can effectively apply them to protect themselves and their peers during experiments. Parents, other family members, and teachers instill in us at a young age a sense of responsibility not only for our own safety but also for the protection of others. Users receive safety instructions from their college professors, and they encourage users to cultivate a safety-conscious culture. Employers, colleagues, managers, and safety professionals can all have something to teach users about maintaining a safe workplace as users progress through their careers. Numerous businesses have implemented stringent safety measures to safeguard not only their users but also their own financial interests, property, and other possessions. It is possible that users' workplace presents the greatest challenge to users' personal safety ethic. This is due to the fact that some companies place little emphasis on safety. As a result, having a safety culture in this industry is one of the most crucial factors.

Why is it imperative to maintain an unwavering commitment to safety? People who lack a strong commitment to safety and have inadequate training in the subject are more likely to place themselves in situations where they or others could be injured. Additionally, those with such a mindset are more likely to place the lives of those around them in danger. And even when users are not at the office, they'll need to adhere to sensible safety practices in order to protect their family and close friends. [95]

The following proclamation serves as the safety ethic's guiding principle: "I work safely, avoid unnecessary risk, and accept responsibility for my own safety." This is a rather simple assumption, but it has far-reaching ramifications and sets out a sequence of necessary steps that must be taken.

To maintain a high level of safety consciousness while working safely, users must be educated in safety, continue learning about safety, comprehend how to detect and evaluate hazards, practice safe procedures, and be trained in safety.

- To avoid placing users in unnecessary danger while working in the laboratory, users must first familiarize themselves with the risks they confront and then acquire the skills necessary to mitigate and manage those risks. [96]

When users embrace responsibility for the safety of others as an act of care, users are accountable for their own safety and the safety of those around them. When users do so, they demonstrate concern for the welfare of others. If the user wishes to complete this endeavor successfully, the user must maintain a constant awareness of not only the user's actions but also those of the other students in the laboratory. In laboratories used to teach undergraduate courses, instructors, teaching assistants, and students share responsibility for maintaining safety; however, their involvement does not relieve users of their own responsibility for safety.

When designing an experiment, it is reasonable to presume that safety considerations have been taken into account. In the context of education, it is easy to become so intent on comprehending the experiment that safety precautions are neglected. This is a potentially hazardous situation. Users' participation in the student safety ethic can necessitate a substantial effort on the users' part. This conclusion is largely because it is reasonable to assume that the experiment was designed with safety in mind. This assumption is the main reason why the situation exists. In other words, it is highly probable that users have been instructed to prioritize the safety of others in everything they do.

This book was developed with the intention of educating users on various safety-related topics over several years, and it was written with that goal in mind. On the other hand, having the proper perspective is an essential component of the student safety ethic. Here is where everything begins.

Safety Rules!

This section concludes with a concise summary of the fundamental safety requirements that apply to laboratory employment. Most of these tips will be clearer as the user reads the rest of the book. Nonetheless, this chapter is a great place to start if users wish to enter a chemistry laboratory without incident and fulfill their responsibilities. [100]

Pay careful attention to both the instructor's and the laboratory's instructions.

Eye protection must be worn at all times when working with chemicals in the laboratory to prevent damage to the eyes. There is a possibility that the user will suffer permanent vision loss if the user does not comply.

The user should always wear apparel that can protect the user from the elements as well as any potential accidents, such as spills. When working in an environment containing chemicals, it is imperative that the user wear gloves that are resistant to those chemicals.

Inside the research facility, users are not permitted to consume, drink, smoke, chew gum, apply cosmetics, or use smokeless tobacco-containing products.

When working with volatile chemicals, flammable liquids or gases, or odorous chemicals, or whenever toxic chemical vapors, particles, or dust may be produced, it is essential to wear a chemical hood. This precaution is done to defend against potentially hazardous substances.

In the laboratory, it is against the rules to engage in disruptive behavior, create an excessive amount of noise (using radios, DVD players, or iPods), or play practical jokes on those who are participating in the experiment.

Users should never put lab chemicals in their mouths, even if told to. Instead, inhale the vapors through the user's nostrils in a gradual and cautious manner. It is not recommended to directly inhale the vapors.

If the user accidentally obtains a chemical on the user's skin or in the user's eyes, the user must immediately wash the affected area with water and notify the user's instructor.

Under no circumstances should the user attempt to manage any laboratory-related responsibilities on their own.

Users are required to provide the instructor with an immediate report of any incidents involving accidents, injuries, leakage, or other potentially hazardous conditions.

Before they can be disposed of in an appropriate fashion, used chemical compounds must be stored for a period of time in the appropriate containers.

Never use a Bunsen burner or any other method that entails an open flame when users need to heat something that can catch fire.

This recommendation applies to all containers used to store solutions or compounds.

The Safety Ethics

If the user needs a refresher on student safety ethics, please refer to the preceding section. This document contains phrases like "I work safely, avoid unnecessary risk, and accept responsibility for safety." When transitioning from an academic to a professional setting, it is essential to update this code of conduct to reflect the additional responsibilities and conditions associated with job status. This is due to the fact that transitioning from self-employment to employment with another party entails an entirely different set of obligations and prerequisites than running one's own business, including the need to adhere to workplace safety regulations and protocols that protect not only oneself but also colleagues and clients.

Why is it imperative to maintain an enduring dedication to safety at all times? The enhancement of users' physical health and the extension of users' life expectancy should be users' highest priorities. People with poor safety ethics and a lack of safety knowledge are more likely to place themselves in perilous situations and have a lower chance of surviving for an extended period of time without causing harm to themselves or others. People with poor safety ethics are more likely to place themselves in potentially hazardous situations. People with a negative attitude toward safety and a limited comprehension of it are more likely to place themselves in dangerous situations. Because of this, there is a greater likelihood that they will endanger the lives of others.

The user should prioritize memorizing the following safety ethic as soon as it is practically possible: "I value safety, work safely, prevent at-risk behavior, promote safety, and accept responsibility for safety." Even though this is a fairly simple premise, it has a large number of consequences and a long list of potential courses of action. Now that this has been established, let's examine each component separately and discuss some interpretive concepts so that we can all agree on what each component represents. [97]

Safety is a value that must never be compromised; safety is an integral part of users' daily activities that requires prudent conduct; and it prevents and shields users, users' family, users' co-workers, and others from injury or suffering. It is a value that should never be compromised; it is an integral part of users' daily activities that necessitates prudent behavior; and it is a value that should never be compromised.

- To work safely requires that user be educated in safety in order to minimize the risks of injury or illness, that user continually strive to learn about safety, that user learn to recognize and evaluate hazards, that user identify and practice safe procedures, that user seek ways to minimize, reduce, and control risks from hazards, and that user maintain a high level of safety awareness at home, at work, and in users' leisure time. If a user wants to prevent risky behavior in themselves and others, the user must first become aware of it, then learn to recognize it when they see it, then refrain from engaging in risky behavior themselves, and finally strive to dissuade others from engaging in risky behavior.
- If the user accepts responsibility for safety, it means that the user is accountable for their own safety, that the user is responsible for the safety of others, and that the user recognizes others for their safe deeds. [98]

Adopting the safety ethic requires a significant effort from the user. For example, to work safely requires that the user learn to recognize and evaluate hazards—and it requires a base of knowledge that the user must build on continuously throughout the user's lifetime, not just in this course of study.

RAMP

In addition to bringing users' prior knowledge of safety to the workplace, users' new position requires them to maintain a positive and assertive attitude towards safety.

This requirement is in addition to users' existing safety knowledge. Users are responsible for conducting an analysis of the risks present at users' new place of employment and determining the most effective method for mitigating them, taking into account the limitations imposed by users' employers' facilities, equipment, and safety precautions. Users are required to assist users' new employer in reducing the risks associated with users' new workplace, and if users are in a position of authority, users also need to assist users' users in learning about safety and its importance so that they can reduce the risks associated with workplace hazards. Users also need to assist users' employers in mitigating the risks associated with users' new workplace.

User must be ready to manage emergencies at users' new workplace and help them prepare by simulating likely scenarios and practicing emergency procedures. In addition, users must be prepared to handle any potential emergencies at their previous workplace. [99]

LAWS AND REGULATIONS PERTAINING TO SAFETY

In order to conduct a portion of an experiment, a college student working in a laboratory required a supply of lithium aluminum hydride [LAH]. After the lid was removed from the can, the LAH inside contacted humid air, causing it to rupture and resulting in burns to the researcher's face and hands. He was transported by an ambulance from the scene to a nearby medical facility. The local fire department was ultimately able to contain the blaze, but there were significant delays before people were allowed to return to the structure. This was due to the fact that the hazardous materials team spent hours ensuring that it was safe for them to proceed. Authorities determined that the facility must remain closed until environmental testing proves beyond a reasonable doubt that it is sufficiently secure to be rented to customers. In addition, the presence of animals used in experiments raised concerns about their welfare on the premises.

Administrative agencies enact statutes and regulations. In the United States, the laws that regulate the functioning of our society are the result of deliberations at the federal, state, and municipal levels of government. These are designed with the well-being of the entire population in mind, and the accompanying documents are frequently exhaustive and elaborate on the desired outcome in outstanding detail. Regulations are a set of rules that are developed to implement the particulars of a law that has already been enacted.

Congress passed the Occupational Safety and Health Act of 1970 in 1970 with the intent of protecting American laborers on the job. The year of the act's ultimate enactment, 1970, inspired its name. Its purpose was to ensure that every man and woman in our country had the opportunity to work in a secure and healthy environment. This statute led to the creation of both the Department of Labor [DOL] and the Occupational Safety and Health Administration [OSHA]. The Department of Labor was charged with bringing the law into effect. The Occupational Safety and Health Administration [OSHA] produced regulations that were subsequently incorporated into the Code of Federal Regulations [CFR] in order to fulfill its responsibility of administering the law. Part 1910 of Title 29 of the Code of Federal Regulations contains the regulations governing the safety and health of the workplace. This code also contains the Department of Labor.

Although there are numerous laws and regulations, we only discuss those that pertain to laboratories. There are times when reading the regulations makes the user feel as though they have entered Dante's Inferno from The Divine Comedy. This feeling is due to the complexity of the regulations. Moreover, given that lobbyists, businesses, unions, and various other political parties are all involved in the rule-making process, it cannot be said that these standards are "only" motivated by scientific considerations. A person cannot use ignorance of the law as an excuse for failing to comply with these standards, and they should make every effort to learn as much as possible about the laws that govern the operation of chemical laboratories to avoid violating these requirements.

SUMMARY

The four fundamental safety concepts, along with the student safety ethic, accentuate the importance of maintaining a secure laboratory environment within the context of the Four Safeguard Principles. The book "Green Chemistry" provides an introduction to the fundamental principles of green chemistry and describes how these concepts can be incorporated into a chemistry curriculum at the college level.

The practice of undertaking chemical reactions in an environmentally responsible manner is known as green chemistry. Regarding the Reconsideration of Safety:

"Laboratory Accidents' Lessons Learned" explains how to draw conclusions and learn lessons from an in-depth analysis of an event by employing the "five whys." The "Environmentally Friendly Chemistry: Inclusion in the Organic Study Program" investigates the applications of environmentally benign chemistry in organic chemistry laboratories. This presentation discusses strategies for developing and promoting a robust safety culture among those operating under users' supervision or leadership as part of the process of establishing a safety culture. The individuals under users' supervision or direction can implement these strategies. Employers have the following expectations for the level of safety knowledge possessed by new chemists:

This paper advises employers on the types of safety skills they should expect newly recruited workers [recent graduates] to possess. The recommendation is presented as a guide. Here, users encounter safety-related regulations and mandates. In this book, which provides a condensed summary of such laws and regulations, the rules and regulations that pertain to chemical and laboratory safety are broken down into their essential components. Environmentally Safe Chemistry: This book discusses several green chemistry concepts that can be utilized to maintain a safe laboratory environment.

REFERENCES

1. Guidelines for Chemical Laboratory Safety in Secondary Schools. (2016). Acs.org. Retrieved July 19, 2023, from https://www.acs.org/content/dam/acsorg/about/governance/committees/chemicalsafety/publications/acs-secondary-safety-guidelines.pdf
2. Res.In. Retrieved July 19, 2023, from https://chem.hbcse.tifr.res.in/wp-content/uploads/2019/03/Safety-rules-in-the-Laboratory-new-PDF.pdf
3. National Research Council (US) Committee on Prudent Practices. (2023). Safety laws and standards pertinent to the laboratory. National Academies Press.
4. Cheremisinoff, N. P., & Rosenfeld, P. E. (2023). Pollution prevention and best practices for the wood-preserving industry. In Handbook of Pollution Prevention and Cleaner Production (pp. 135–178). Elsevier.
5. Prohibition against eating and drinking in labs. (2023). Uiowa.edu. Retrieved July 19, 2023, from https://ehs.research.uiowa.edu/lab-safety/prohibition-against-eating-and-drinking-labs
6. United States Congress House of Representatives. (2023). To amend the Public Health Service Act to provide support for the modernization and construction of biomedical and behavioral research facilities and laboratory instrumentation. Bibliogov.
7. Safe lab practices. (2023). Ucsc.edu. Retrieved July 19, 2023, from https://ehs.ucsc.edu/programs/research-safety/safe-lab-practices.html
8. Koti Reddy, B., Ayyagari, K. S., & Medam, R. R. (2022). Electrical safety for residential and rural microgrids. In Residential Microgrids and Rural Electrifications (pp. 293–320). Elsevier.
9. Emergency safety equipment. (2023). Uoregon.edu. Retrieved July 19, 2023, from https://safety.uoregon.edu/emergency-safety-equipment
10. StackPath. (2023). Mlo-online.com. Retrieved July 19, 2023, from https://www.mlo-online.com/continuing-education/article/13017006/keeping-the-laboratory-environment-clean-and-safe
11. BLiSC (NCBS, inStem, C-CAMP) responses to COVID-19. (2023). Res.In. Retrieved July 19, 2023, from https://www.ccamp.res.in/blisc-ncbs-instem-c-camp-responses-covid-19
12. Apa.org. Retrieved July 19, 2023, from https://www.apa.org/science/leadership/care/guidelines
13. Laboratory safety rules and regulations. (2023). Webassign.net. Retrieved July 19, 2023, from https://www.webassign.net/ncsumeorgchem1/frontMatter/saftey/saftey.html
14. Grossel, S. S. (2023). Reduce hazards in process vacuum systems: reduce explosion risks, and chemical and physical hazards, to ensure safer operation of vacuum pumps and related systems. Chemical Engineering, 119, 47+. https://go.gale.com/ps/i.do?p=AONE&u=googlescholar&id=GALE|A306862138&v=2.1&it=r&sid=AONE&asid=f692333d
15. Lisa Nichols, Butte College. (2017, November 20). 5.6A: Overview of rotary evaporation. Chemistry LibreTexts; Libretexts. https://chem.libretexts.org/Bookshelves/Organic_Chemistry/

Organic_Chemistry_Lab_Techniques_(Nichols)/05%3A_Distillation/5.06%3A_Rotary_Evaporation/5.6A%3A_Overview_of_Rotary_Evaporation

16. Pressure and vacuum systems. (2023). Office of Environmental Health and Safety. Retrieved July 19, 2023, from https://ehs.princeton.edu/laboratory-research/laboratory-safety/laboratory-equipment-and-engineering/pressure-and-vacuum-systems

17. Sigmaaldrich.com. Retrieved July 19, 2023, from https://www.sigmaaldrich.com/deepweb/assets/sigmaaldrich/marketing/global/documents/404/391/glass_care_safe_handling.pdf

18. National Research Council (US) Committee on Prudent Practices. (2023). Working with laboratory equipment. National Academies Press.

19. Lab Safety Guide. (2023). Stonybrook.edu. Retrieved July 19, 2023, from https://ehs.stonybrook.edu/programs/laboratory-safety/lab-safety-guide

20. Raunek. (2021, January 20). Boiler operation made easy: Procedure for starting and stopping a boiler. Marine Insight. https://www.marineinsight.com/tech/boiler/boiler-operation-made-easy-procedure-for-starting-and-stopping-a-boiler/

21. National Research Council (US) Committee on Prudent Practices. (2023). Management of waste. National Academies Press.

22. Safety in the chemistry laboratory. (2023). Cerritos.edu. Retrieved July 19, 2023, from https://www.cerritos.edu/chemistry/_includes/docs/General/cerritos_Safety_in_the_Chem_Lab_June%2011%20%20%20-09.pdf

23. Udel.edu. Retrieved July 20, 2023, from http://www1.udel.edu/chem/white/C601/TA-Tales.pdf

24. Hazard and risk - general. (2023, June 13). Ccohs.Ca. https://www.ccohs.ca/oshanswers/hsprograms/hazard/hazard_risk.html

25. About exposure. (2023). Health.ny.gov. Retrieved July 20, 2023, from https://www.health.ny.gov/environmental/about/exposure.htm

26. (2023). Cartercenter.org. https://www.cartercenter.org/resources/pdfs/health/ephti/library/lecture-notes/env-occupational-health-students/ln-occ-health-safety-final.pdf

27. Osha.gov. Retrieved July 20, 2023, from https://www.osha.gov/sites/default/files/publications/OSHA3404laboratory-safety-guidance.pdf

28. Chemistry laboratory safety specifications. (2023). CT.Gov - Connecticut's Official State Website. Retrieved July 20, 2023, from https://portal.ct.gov/SDE/Publications/Connecticut-High-School-Science-Safety/Chemistry-Laboratory-Safety-Specifications

29. Laboratory Safety Rules. (2023). In Environmental Sampling and Analysis for Metals (pp. 336–343). CRC Press.

30. International Labor Office, & Alli, B. O. (2023). Fundamental principles of occupational health and safety (2nd ed.). International Labor Office.

31. Abdalla, S., Apramian, S. S., Cantley, L. F., & Cullen, M. R. (2023). Occupation and Risk for Injuries. In Disease Control Priorities, Third Edition (Volume 7): Injury Prevention and Environmental Health (pp. 97–132). The World Bank.

32. National Research Council (US) Committee on Hazardous Biological Substances. (2023). Safety management. National Academies Press.

33. StackPath. (2023). Mlo-online.com. Retrieved July 20, 2023, from https://www.mlo-online.com/management/lab-safety/article/21250809/laboratory-safety-management-roles-and-responsibilities

34. Nationalacademies.org. Retrieved July 20, 2023, from https://nap.nationalacademies.org/resource/21918/Chemical-Laboratory-Safety-and-Security.pdf

35. Lim, P. J. Y., Heindel, N. D., Engida, T., Ababa, A., El-Khateeb, M., & Thulstrup, E. W. (2023). International union of pure and applied chemistry the academy of sciences for the developing world. Nationalacademies.org. Retrieved July 20, 2023, from https://nap.nationalacademies.org/resource/21918/execsummary-brochure-English.pdf

36. National Research Council (US) Committee on Prudent Practices. (2023). Environmental health and safety management system. National Academies Press.

37. Laboratory Supervisor Safety Responsibilities Policy. (2023). University of Nevada, Reno. Retrieved July 20, 2023, from https://www.unr.edu/ehs/policies-manuals/laboratory-supervisors

38. Laboratory Supervisor. (2023). Petrochemworks.com. Retrieved July 20, 2023, from https://petrochemworks.com/job/laboratory-supervisor

39. Oecd.org. Retrieved July 20, 2023, from https://www.oecd.org/education/school/37133393.pdf

40. Safety, R. in. (2023). Teacher and Student. Flinnsci.com. Retrieved July 20, 2023, from https://www.flinnsci.com/api/library/Download/5d960d6553744e4393c35055cf244089

41. Nationalacademies.org. Retrieved July 20, 2023, from https://nap.nationalacademies.org/resource/21918/Chemical-Laboratory-Safety-and-Security.pdf

42. Deepak. (2018, September 24). Prevention of accidents in laboratory stores. Lab-training.com. https://lab-training.com/prevention-of-accidents-in-laboratory-stores/

43. Osha.gov. Retrieved July 20, 2023, from https://www.osha.gov/sites/default/files/publications/OSHA3404laboratory-safety-guidance.pdf

44. Laidlaw, L. (2022, March 13). Laboratory eye protection. Safety Partners, Inc. https://www.safetypartnersinc.com/laboratory-eye-protection/

45. Uwyo.edu. Retrieved July 20, 2023, from https://www.uwyo.edu/safety/chemical/docs/lab_attire_pub.pdf

46. Selection and use of gloves in the laboratory. (2023). Dartmouth.edu. Retrieved July 20, 2023, from https://www.dartmouth.edu/ehs/chemical/chemical_gloves.html

47. National Research Council (US) Committee on Prudent Practices. (2023). Evaluating hazards and assessing risks in the laboratory. National Academies Press.

48. Carruth, G., & Ehrlich, E. (2023). The Giant Book of American Quotations. Knowledge, 3.

49. Thompson, J. S., & Edmonds, O. P. (2023). Safety aspects of handling the potent allergen FDNB. The Annals of Occupational Hygiene, 23(1), 27–33. https://doi.org/10.1093/annhyg/23.1.27

50. Cas.org. Retrieved July 7, 2023, from http://www.cas.org/expertise/cascontent/

51. Osha.gov. Retrieved July 7, 2023, from http://www.osha.gov/pls/oshaweb/

52. Phys. Retrieved July 7, 2023, from http://jrm.phys.

53. Where to find MSDS and SDS on the Internet. (2023). Ilpi.com. Retrieved July 7, 2023, from http://www.ilpi.com/msds/index.html

54. Directorate of standards and guidance - directory of offices & staff. (2023). Osha.gov. Retrieved July 7, 2023, from http://www.osha.gov/dsg/

55. American National Standards Institute (ANSI). Z400.1, Hazardous Industrial Chemicals-Material Safety Data Sheets- Preparation. (2023). ANSI.

56. Globally Harmonized System of Classification and Labelling of Chemicals (GHS), 2nd revision, United Nations. (2023).

57. Robert, J., & Alaimo, L. A. (2019). Material Safety Data Sheets. In: Handbook of Chemical Health and Safety (R. J. Alaimo, Ed.). American Chemical Society Oxford University Press.

58. crisis communication. (2023). for the Public Health. Who.int. Retrieved July 20, 2023, from https://apps.who.int/iris/rest/bitstreams/52481/retrieve

59. Jefferson, T. (2019). Excerpt from notes on the state of Virginia. In P. C. Mancall (Ed.), Land of Rivers (pp. 55–61). Cornell University Press.

60. Lessons learned at UC Berkeley. (2023). Berkeley.edu. Retrieved July 7, 2023, from http://www.ehs.berkeley.edu/lessonslearned.html

61. Homepage. (2023). Unece.org. Retrieved July 7, 2023, from http://www.unece.org/

62. Mali, R. GHS hazard classification: Everything user need to know. Era-environmental.com. https://www.era-environmental.com/blog/ghs-hazard-classification

63. Homepage. (2023). Unece.org. Retrieved July 7, 2023, from http://www.unece.org/

64. Hazard communication - overview. Osha.gov. Retrieved July 7, 2023, from http://www.osha.gov/dsg/hazcom/ghs.html

65. Ali, A. A., Shaalan, N., Al-Dahhan, W., & Usersif, E. (2023). For a safer working environment with hydrofluoric acid in Iraqi industrial plants. Open Journal of Safety Science and Technology, 06(04), 77–80. https://doi.org/10.4236/ojsst.2016.64007

66. Ali, A. A., Shaalan, N., Al-Dahhan, W., & Usersif, E. (2023). For a safer working environment with hydrofluoric acid in Iraqi industrial plants. Open Journal of Safety Science and Technology, 06(04), 77–80. https://doi.org/10.4236/ojsst.2016.64007

67. Rasool, S. R., Al-Dahhan, W. H., Al-Zuhairi, A. J., Hussein, F. H., Rodda, K. E., & Usersif, E. (2023). Fire and Explosion Hazards expected in a Laboratory. Journal of Laboratory Chemical Education, 4(2), 35–37.

68. Staehle, I. O., Chung, T. S., Stopin, A., Vadehra, G. S., Hsieh, S. I., Gibson, J. H., & Garcia-Garibay, M. A. (2023). An approach to enhance the safety culture of an academic chemistry research laboratory by addressing behavioral factors. Journal of Chemical Education, 93(2), 217–222. https://doi.org/10.1021/acs.jchemed.5b00299

69. Al-Dahhan, W. H., Al-Zuhairi, A. J., Hussein, F. H., Rodda, K. E., & Usersif, E. (2023). Laboratory biological safety cabinet (BSC) explosion. Karbala International Journal of Modern Science, 2(4), 276–279. https://doi.org/10.1016/j.kijoms.2016.11.001

70. Clinical Laboratory Standards Institute, document GP18-A2. Laboratory Design.

71. Gile, T. J. (2019). Ergonomics bring safety features to an aging laboratory workforce. MLO: Medical Laboratory Observer, 41(1), 26–27.

72. Robert, J. A. Handbook of Chemical Health and Safety. American Chemical Society Handbooks.

73. Al-Ibadi, M. (2023). The Chemical Safety and Security in the University of Kufa: Progress and Challenges. Iraqi National Journal of Chemistry, 16(2), 96–105.

74. Al-Khalaf A, Green Chemistry and Sustainability of Chemical Safety and Security, Iraqi National Journal of Chemistry, 16(2), 71-81, (2023).

75. Kareem, S., Hamzah, I., & Bahiya, H. (2023). Improve Biosafety Level in Advance Molecular Laboratory at College of Science. Iraqi National Journal of Chemistry, 16(2), 89–95.

76. Al-Zuhairi, A., Al-Dahhan, W., Hussein, F., Rodda, K., & Usersif, E. (2023). A Vision to Promote the Forensic DNA Facility at Al-Nahrain University in Terms of Safety Measures. Oriental Journal of Physical Sciences, 2.

77. Toxicity and hazard exposure. (n.d.). Uccs.edu. Retrieved July 20, 2023, from https://ehs.uccs.edu/sites/g/files/kjihxj1296/files/inline-files/UCCS.HMMP%20Attachment%20C%20-%20Toxicity%20and%20Hazard%20Exposure.pdf

78. Wikipedia contributors. (2023, May 20). Biological hazard. Wikipedia, The Free Encyclopedia. https://en.wikipedia.org/w/index.php?title=Biological_hazard&oldid=1156044563

79. Ttu.edu. Retrieved July 20, 2023, from https://www.depts.ttu.edu/che/Lab_Safety_Manual_MAY2018.pdf

80. Recommended safety and emergency equipment for the laboratory. (2021, June 18). Carolina Knowledge Center. https://knowledge.carolina.com/professional-growth/safety/recommended-safety-and-emergency-equipment-for-the-laboratory/

81. Fontalvo, C. (2020, October 20). Fume Hoods, their use and functions of in lab environment. Lab Bubble. https://lab-bubble.com/2020/10/20/fume-hoods-their-use-and-functions-of-in-lab--environment/

82. United States Congress House of Representatives. Transitioning the environmental measurements laboratory to the Department of Homeland Security. Bibliogov.

83. Myers, K. Compressed gas cylinders. Lab Manager; Lab Manager Magazine. https://www.labmanager.com/compressed-gas-cylinders-7295

84. Laboratory chemical waste management guidelines. (2023). Upenn.edu. https://ehrs.upenn.edu/sites/default/files/2018-02/wastemanual2017final.pdf

85. Dahal, P. (2023, March 21). 52 laboratory safety symbols, signs, and meanings. The Biology Notes. https://thebiologynotes.com/laboratory-safety-symbols/

86. Das, A. (2021, August 16). 12 common lab safety signs: Keeping safe in the lab. Bitesize Bio; Science Squared Ltd. https://bitesizebio.com/28135/safety-comes-first-common-laboratory-signs/

87. McLeod, V. Laboratory hazards and risks. Lab Manager; Lab Manager Magazine. https://www.labmanager.com/laboratory-hazards-and-risks-18238

88. Ali, H. (2023, April 1). Laboratory hazards and safety controls. Occupational Health and Safety Blog; Health & Safety. https://www.hseblog.com/laboratory-hazards-safety-controls-lab-safety/

89. Electrical hazards. (2019, June 18). Laboratory Safety Manual; Alfaisal University | Laboratory Safety Manual. https://lsm.alfaisal.edu/documentations/electrical-hazards/

90. Laboratory Housekeeping. (2023). Cloudfront.net. Retrieved July 21, 2023, from https://de065k6q0yws1.cloudfront.net/wp-content/uploads/sites/25/2017/01/LaboratoryHousekeeping.pdf

91. Food and beverages in the lab. (2023). Fsu.edu. Retrieved July 21, 2023, from https://safety.fsu.edu/safety_manual/Food%20and%20Beverages%20in%20the%20Lab.pdf

92. Lab safety equipment every laboratory should have. (2023). Workstationindustries.com. Retrieved July 21, 2023, from https://resources.workstationindustries.com/blog/lab-safety-equipment

93. Lindhout, P., & Reniers, G. (2021). Involving moral and ethical principles in safety management systems. International Journal of Environmental Research and Public Health, 18(16), 8511. https://doi.org/10.3390/ijerph18168511

94. Laboratory safety principles. (2023). Laboratory Safety Principles. Retrieved July 21, 2023, from https://www.wcu.edu/discover/campus-services-and-operations/facilities-management/safety-and-risk-management/laboratory-safety/laboratory-safety-principles.aspx

95. Carter, E. Responsible conduct of research. Cuanschutz.edu. Retrieved July 21, 2023, from https://research.cuanschutz.edu/docs/librariesprovider148/crsc_documents/education-materials/rcr-seminar-materials/rcr6_safe-laboratory-practice-and-laboratory-tools-class-slides.pdf?sfvrsn=d42882b9_6

96. Prudent practices in the laboratory: Handling and management of chemical hazards, updated version, Committee on Prudent Practices in the Laboratory: An Update; National Research Council. (2023). Cwu.edu. Retrieved July 21, 2023, from https://www.cwu.edu/ehs/sites/cts.cwu.edu.ehs/files/documents/Prudent%20Practices%20in%20the%20Laboratory.pdf

97. Mehlich, J., Moser, F., Van Tiggelen, B., Campanella, L., & Hopf, H. (2023). The ethical and social dimensions of chemistry: Reflections, considerations, and clarifications. Chemistry (Weinheim an Der Bergstrasse, Germany), 23(6), 1210–1218. https://doi.org/10.1002/chem.201605259

98. Bellarmine.edu. Retrieved July 21, 2023, from https://www.bellarmine.edu/docs/default-source/chemistry-docs/safety-plan-update-2018.pdf?sfvrsn=dd1b6880_0

99. How can occupational safety and health be managed? (Labor administration and inspection). Retrieved July 21, 2023, from https://www.ilo.org/global/topics/labour-administration-inspection/resources-library/publications/guide-for-labour-inspectors/how-can-osh-be-managed/lang--en/index.htm

100. Manager, L. (2023). Lab safety rules and guidelines. Lab Manager; Lab Manager Magazine. https://www.labmanager.com/science-laboratory-safety-rules-guidelines-5727

www.ingramcontent.com/pod-product-compliance
Lightning Source LLC
LaVergne TN
LVHW060043210926
848481LV00013B/3945
* 9 7 9 8 9 0 4 3 1 3 6 0 9 *